GEEK HERESY

GEEK HERESY

RESCUING SOCIAL CHANGE FROM THE CULT OF TECHNOLOGY

KENTARO TOYAMA

PublicAffairs • New York

Published in the United States by PublicAffairs™,
a Member of the Perseus Books Group

Printed in the United States of America.

PublicAffairs books are available at special discounts for bulk purchases in the U.S. by corporations, institutions, and other organizations. For more information, please contact the Special Markets Department at the Perseus Books Group, 2300 Chestnut Street, Suite 200, Philadelphia, PA 19103, call (800) 810–4145, ext. 5000,
or e-mail special.markets@perseusbooks.com.

Book Design by Cynthia Young
Typeset in 11.5 Adobe Garamond Pro

The Library of Congress has cataloged the printed edition as follows: [TK]

ISBN 978–1-61039–528–1 (HC)

ISBN 978–1-61039–529–8 (EB)

First Edition

10 9 8 7 6 5 4 3 2 1

To my mother,
who nurtured what heart, mind, and will I have.

To my father,
who taught me to follow my aspirations.

To Rohan,
for whom I wish much intrinsic growth.

And to Bill Gates, cofounder of Microsoft
and the Bill & Melinda Gates Foundation,
both of which gave me opportunities to learn
what I write about in this book.

Gates once wrote, "The first rule of any technology used in a business is that automation applied to an efficient operation will magnify the efficiency. The second is that automation applied to an inefficient operation will magnify the inefficiency." (The Road Ahead, 1995)

CONTENTS

Introduction ix

Part 1

CHAPTER 1: No Laptop Left Behind 3
Conflicting Results in Educational Technology

CHAPTER 2: The Law of Amplification 17
A Simple But Powerful Theory of Technology's Social Impact

CHAPTER 3: Geek Myths Debunked 38
Dispelling Misguided Beliefs about Technology

CHAPTER 4: Shrink-Wrapped Quick Fixes 57
Technology as an Exemplar of the Packaged Intervention

CHAPTER 5: Technocratic Orthodoxy 74
The Pervasive Biases of Modern Do-Gooding

Part 2

CHAPTER 6: Amplifying People 103
The Importance of Heart, Mind, and Will

CHAPTER 7: A Different Kind of Upgrade 122
Human Development Before Technology Development

CHAPTER 8: Hierarchy of Aspirations 151
The Evolution of Intrinsic Motivation

CHAPTER 9: "Gross National Wisdom" 172
Societal Development and Mass Intrinsic Growth

CHAPTER 10: Nurturing Change 192
Mentorship as a Social-Cause Paradigm

Conclusion 211

Acknowledgments 219
Appendix: Highlighted Nonprofits 223
Notes 225
References 279
Index 319

INTRODUCTION

"Talent is universal; opportunity is not." That's how Megan Smith, chief technology officer of the United States and former vice president of Google.org, began her opening remarks at the University of California, Berkeley, in the spring of 2011. She and I were on a panel titled "Digital Divide or Digital Bridge: Can Information Technology Alleviate Poverty?"[1] The event was held in South Hall, the campus's oldest building but home to its youngest school–the School of Information–where scholars study the interaction between digital technology and human society. The hall was packed. The panel drew not only students and faculty, but also Bay Area impact investors, nonprofit leaders, and social entrepreneurs.

Google.org's motto at the time was "tech-driven philanthropy," and Smith embraced it.[2] She implicitly agreed that talent was universal. But, she said, "opportunity is becoming more universal" as well. According to Smith, opportunity was expanding along with "the network," by which she meant the Internet, mobile phone systems, and presumably the Google technologies riding on them.

That more people are becoming connected is a fact. By the end of 2014, there were nearly 3 billion people on the Internet. Sometime in 2015 the total number of mobile phone accounts will exceed the world population.[3] Both figures continue to grow. Smith suggested that these technologies bring people together, trigger revolutions, and make "all world knowledge . . . available online for free." If she is right, everyone everywhere will soon have plenty of opportunity: Talent is universal, and opportunity is the Internet.

The world's leading technologists thoroughly agree, and they're competing to speed things up. In 2009, Sir Tim Berners-Lee, the inventor of the key protocols that drive the Internet, founded the World Wide Web Foundation to spread the Web as "a global public good and a basic right." Its tagline: "Connecting People. Empowering Humanity."[4] A couple years later, Smith's colleagues at Google began working to deliver WiFi through solar-powered balloons. CEO Larry Page says, "Two out of three people in the world don't have good Internet access now. We actually think [balloon-delivered Internet] can really help people."[5] Not to be outdone, Facebook founder Mark Zuckerberg announced Internet.org in 2013. "We've been working on ways to beam internet to people from the sky," he posted.[6] He wants to reach remote places with infrared lasers and high-altitude drones.

That tech giants are messianic about their creations is no surprise. But their outlook has possessed powerful people outside of Silicon Valley, too. US Secretary of Education Arne Duncan said that "technology is a game-changer in the field of education–a game-changer we desperately need to both improve achievement for all and increase equity for children and communities who have been historically underserved."[7] Economist Jeffrey Sachs, author of *The End of Poverty* and the force behind the United Nations' Millennium Villages Project, believes that "mobile phones and wireless Internet end isolation, and will therefore prove to be the most transformative technology of economic development of our time."[8] And in 2011, then–secretary of state Hillary Clinton announced a new foreign policy doctrine. She introduced "Internet freedom" by saying that information networks were a "great leveler" that we should use "to help lift people out of poverty and give them a freedom from want."[9] World leaders are convinced that technology will make the world a better place.

But does technology really cause positive social change?

Consider poverty in the United States. Its rate decreased steadily for decades until 1970. Around 1970, though, the decline stopped. Since then, the poverty rate has held steady at a stubborn 12 to

13 percent–embarrassingly high for the world's richest country–only to rise since the 2007 recession.[10] Over the past four decades, real incomes for poor and middle-class households stagnated. Inequality shot up to a level not seen for a century.[11]

During the same four decades, though, the United States experienced an explosion of new technologies. America ushered in the Internet and the personal computer. Its companies invented mobile phones and social media. US firms such as Apple, Google, Microsoft, and Twitter dominate the global corporate landscape, churning out product after product used by billions of people. In 2014 there were more than 210 million Facebook accounts in the United States, outnumbering Americans aged fifteen to sixty-four. For a while now, the total US population has been eclipsed by the number of wireless subscriptions.[12]

So during a golden age of innovation in the world's most technologically advanced country, there has been no dent in our rate of poverty.[13] All of our amazing digital technologies, widely disseminated, didn't alleviate our most glaring social ill.

A Tale of Two Approaches

When Smith said, "Talent is universal; opportunity is not," she was quoting an epigraph from a memoir, *It Happened on the Way to War*, by former Marine captain Rye Barcott. Barcott was an officer-in-training in 2000 when he visited Kibera, the largest slum in Nairobi, and his eyes were opened to global poverty. Feeling compelled to do something about it, he worked with local residents Tabitha Atieno Festo and Salim Mohamed to found a nonprofit organization called Carolina for Kibera (CFK), which has since been honored for its work by *Time* magazine and the Bill & Melinda Gates Foundation. The organization runs health and education programs and trains youth leaders to solve community problems. Steve Juma, for example, joined a CFK youth soccer team and discovered he made a good referee and peer mentor. CFK granted him a medical school scholarship, and Juma now treats patients at its clinic.[14] The founders believe that everyone comes into the world with potential, but not everyone has the opportunity to develop it. Talent is universal, and opportunity is the nurturing of that talent.

This is very different from Smith's take. Whereas Smith is concerned with external provision, Barcott builds up internal strengths. The difference is profound, as even Smith–and every Google employee–knows well from another context. Google posts its job announcements on its website, so in theory, anyone with access to the Internet has the opportunity to apply. In practice, though, the jobs are closed to all but a small minority of people who have the education, experience, and personal contacts to pass extensive rounds of interviews and aptitude tests. I know many low-income people who would like nothing more than a well-paying job at a global technology company. But it doesn't matter whether they can browse engineering jobs on their phones. Online opportunity isn't always actual opportunity.

Of course, Smith herself wouldn't argue that a tenth-grade dropout from East Palo Alto and a Stanford computer science PhD have the same ability to hack software. Nevertheless, by equating the Internet with opportunity for underprivileged people, she has made a dubious assumption–an assumption that the Internet can make up for severe non-Internet deficiencies.

So–talent is universal; opportunity is not. The same six words capture what Smith and Barcott both believe. Yet their different interpretations lead to wildly divergent ways of trying to change the world. Smith wants to spread technology to every corner of the planet. Barcott focuses on cultivating individual talents. One builds technologies. The other fosters people.

I know very well where Smith was coming from. For twelve years I worked at Microsoft, where, like every other gizmo-happy technologist, I unconsciously embraced a peculiar paradox. It revealed itself in the most innocuous things that the company said. At corporate gatherings, executives would tell us, "You are our greatest asset!" But in their marketing, they would tell customers, "Our technology is your greatest asset!" In other words, what matters most to the company is capable people, but what should matter to the rest of the world is new technology. Somehow what was best for us and what was best for others were two different things.[15]

This book is about this subtle contradiction and its outsize consequences. I explore how a misunderstanding about technology's role in society has infected us–not just the tech industry, but global civilization as a whole–and how it confuses our attempts to address the world's persistent social problems. The confusion expresses itself as Silicon Valley executives who evangelize cutting-edge technologies at work but send their children to Waldorf schools that ban electronics. Or as a government that spies on its citizens' emails while promoting the Internet abroad as a bulwark of human rights. Or as a country densely crisscrossed with interactive social media that is nevertheless more politically polarized than ever. *Geek Heresy* demystifies these contradictions and seeks to illuminate a more effective path to social change.

Technoholics Anonymous

I am a recovering technoholic. I was once addicted to a technological way of solving problems.

My parents were nerds at heart, possibly reflecting the stereotypical Japanese fascination with science and technology. On birthdays, they gave me Lego blocks and Erector sets. I have fond memories of playing with a clever Japanese toy called *Denshi Burokku*. It consisted of analog electronics embedded in plastic cubes that you could arrange and rearrange to build lie detectors and radios.[16] By the fifth grade, I was programming an Apple II personal computer. My bookshelves were filled with biographies of Isaac Newton, Thomas Edison, and the Wright brothers as well as titles such as *How Things Work* and *Tell Me Why*.

One book that left a deep impression on me described Russian efforts to build a fusion reactor. When I was growing up in the 1970s, a series of energy crises caused long lines at gas stations and an adult obsession with turning off lights. These developments seemed connected to world events that caused furrows in President Jimmy Carter's brow every time he appeared on TV. Nuclear fusion–as a source of unlimited energy–seemed like it could put an end to these problems once and for all. I thought I could help make it work.

So in college I majored in physics, but, as often happens, one thing led to another, and I changed fields. I did a PhD in computer science, and after that, I took a job at Microsoft Research–one of the world's largest computer science laboratories. What didn't change was my search for technological solutions.

At first I worked in an area called computer vision, which tries to give machines a skill that one-year-olds take for granted but that science still toils to explain: converting an array of color into meaning–a crib, a mother's smile, a looming bottle. Computers still can't recognize these objects reliably, but the field has made progress. For example, these days we don't think twice about the little squares that track a person's face on our mobile-phone cameras. That's a technology that a colleague of mine developed just fifteen years ago.[17] In my own research, I worked on algorithms that allowed you to cut out objects in digital photographs and automatically fill in the hole with an appropriate background.[18] Another project was a precursor to the software in Microsoft's Kinect system, which does away with joysticks for Xbox games by tracking players' physical movements with cameras.[19]

These advances were exciting. They proved the incredible power of technology. And they kept me engaged for seven years. But I began to feel a little dissatisfied with the kind of impact I was having. If I was overly ambitious as a child to think that I could solve big energy problems, now I didn't feel ambitious enough. I wanted to do more than serve the world's gadget lovers.

So in 2004, when my manager in Redmond asked me if I would join him to launch a research center in India–what would become Microsoft's only major lab in the developing world–I jumped at the chance.[20] I was excited by a new topic: How could electronic technologies contribute to social causes in the world's poorest communities? Within months, I moved to Bangalore, expecting to spend a few years applying my technical skills to a new set of problems. What I didn't foresee was that India would change my entire conception of technology.

Prolegomenon

America is a bubble where everyone tunes into YouTube; Amazon delivers to every Kindle; and debates of fact are quickly settled by consulting our iPhones. As a result, it's hard to gain a true sense for technology's real effect on society. We're all breathing the same air. We have a perspective that some people call WEIRD–Western, Educated, Industrialized, Rich, and Democratic.[21]

Outside our bubble, there are places like India–an ocean of diversity presenting a Technicolor range of man-machine interaction. Nonliterate dollar-a-day rickshaw drivers, who are savvy users of Bluetooth file exchange on their multimedia phones, deliver undergraduate computer-science majors to campuses in which programming is taught entirely on paper. Inadequate theories of technology don't hold together in the rough waters of such contrasts.

In Part 1, I'll share what I learned in India and other places about digital technology and demonstrate that the lessons apply everywhere. I'll describe the Law of Amplification, which concisely explains technology's impact on society and shatters pervasive myths about social change. These myths are embedded deep in the modern technocratic psyche, and they mislead us toward mirages that vanish on closer inspection. Part 1 will provoke tech optimists, vindicate tech skeptics, and liberate others from the cult-like hold of technology.

Part 2 suggests the path forward. It will reveal rules for the best ways to apply technology, but move beyond machines to highlight the critical role of individual and societal intention, discernment, and self-control. I'll tell moving stories of extraordinary people, such as Patrick Awuah, a Microsoft millionaire who left his lucrative engineering job to open Ghana's first liberal arts university, and Tara Sreenivasa, a graduate of a remarkable South Indian school that takes children from dollar-a-day households into the high-tech offices of Goldman Sachs and Mercedes-Benz. Part 2 reanimates an ancient narrative for progress that is more relevant today than ever before: Even in a world of abundant technology, there is no social change without change in people.

Throughout, I use examples from global poverty to represent a range of societal afflictions. In part, this is because of my own focus over the past decade. But poverty is also linked to just about every social problem, either directly or by analogy. Being poor often means having lower levels of health, education, and political power. Resource scarcity and environmental destruction are everyday facts in impoverished communities. All forms of social inequality and prejudice echo the motifs of economic inequality and discrimination. By the end of this book, I hope you'll agree that the Law of Amplification and the case for certain human values apply not just to the alleviation of poverty, but to any kind of positive social change.

Greek Geek

In Greek mythology, Daedalus was a brilliant craftsman and engineer. He designed the labyrinth that contained the Minotaur. He devised new methods of carpentry and shipbuilding. His animated statues were the world's first robots.[22] But Daedalus is perhaps best known for the invention of flight. Imprisoned in a tower with his son, Icarus, Daedalus fashioned wings out of feathers and wax. As they planned their escape, he warned Icarus not to fly too close to the sun for fear that the wax would melt. Once they were in the air, though, Icarus ignored his father's warnings. He soared exuberantly into the sky. His wings fell apart, and Icarus fell to his death.

This story is often interpreted with a moral for children: Obey your parents. Reign in hubris. But there is also a timeless lesson for the grownups: Brilliant technology is not enough to save us from ourselves. Tech proponents will insist that Daedalus needed wings to escape. Tech skeptics will say that Icarus would have been better off without them. But had Icarus exercised restraint, or had Daedalus taken more time with his son to convey the risks, they could have benefited from the technology without the tragedy. The real lesson, then, is not about technology at all–it's about the right kind of heart, mind, and will.

PART 1

No Laptop Left Behind

CHAPTER 1

No Laptop Left Behind

Conflicting Results in Educational Technology

India has pole-vaulted onto the global stage as an IT superpower, but only a thin stratum of the country's educated elite is a part of that phenomenon. The rest–as many as 800 million people who live on less than two dollars a day–are lucky if they can work as servants to the rising middle class. There is a cavernous skills gap. Within the glass pyramids and shiny domes of Bangalore's tech acropolises, recruiters struggle to find qualified engineers in a country with four times the population of the United States. Large IT firms like Infosys are so desperate for technical talent that they hire history majors on the basis of IQ tests and then put them through five-month courses in computer programming. Every year, more than 20 million Indians turn twenty years old, yet too few receive the foundational education required to fill the several hundred thousand technical jobs that corporations post each year.

In a country brimming with information technology but lacking in basic education, it seemed natural to investigate how personal computers could support learning. So, that was one of the things I focused on when I moved to India in 2004. I hired a team of designers, engineers, and social scientists. We began projects in education, agriculture, health care, governance, microfinance, and so on. For education, we started by spending time in rural India's government schools. They were blighted by absent teachers, broken toilets, and unquestioning parents.

Desperate administrators often turned to technology as a solution. A startling number of rural schools had computer labs. Because of small budgets, though, the labs were limited to a handful of PCs. Joyojeet Pal, one of our first interns, visited twelve schools across four states and returned with photo after photo of students piled on like rugby players around a single PC.[1] There were never enough terminals for all the children. One dominant child–often an upper-caste boy–tended to monopolize the mouse and keyboard while others crowded around, hoping to have a chance to interact.

It was a perfect opportunity for innovation: What if we plugged in multiple mice per computer, each with a corresponding cursor on screen? As with a video game console, many children could engage simultaneously. Udai Singh Pawar, a smart young researcher in my group with a boyish sense of fun, ran with the idea. He quickly prototyped what we called MultiPoint, along with its own educational software.

Students loved it, and formal experiments confirmed its effectiveness. Pawar verified that for activities like vocabulary drills, students learned just as much with MultiPoint as with a single PC all to themselves.[2] One child, enthralled with the prototype, asked, "Why doesn't every computer come with multiple mice?" We filed a patent, convinced Microsoft to release a free software development kit, and imagined that schools around the world would benefit. We declared victory, and temporarily forgot about the lack of toilets, the silent parents, and the absent teachers.

Projects such as MultiPoint won us awards and recognition. Children inevitably smiled in front of new technology, and politicians loved photo-ops where they handed out new gadgets. I often found myself in teak-paneled boardrooms discussing technology strategy with government ministers, World Bank officials, and nonprofit luminaries. Our research seemed to offer proof that there were technological solutions to developing-world education.

We were not alone. An even bigger splash was made by a nonprofit with an ambitious name that reflected its ambitious plan: One Laptop Per Child. Led by Massachusetts Institute of Technology (MIT) Media Lab founder Nicholas Negroponte, the organization sought to design a $100 laptop that could be sold to developing countries a million units at a time. At the 2005 World Summit on the Information Society, Negroponte shared the stage with United Nations Secretary-General Kofi Annan. They unveiled what looked like a green-and-white Fisher-Price toy that boasted a fully operational PC with kid-friendly software. Annan gave an unabashed endorsement: "These robust and versatile machines will enable kids to become more active in their own learning."[3]

Negroponte summed up our credo as technologists: "It's not a laptop project. It's an education project."[4] By inventing and disseminating new, low-cost devices for learning, we believed we were improving education for the world's less privileged children. But were we?

Geeks Bearing Gifts

The success of our MultiPoint trials encouraged us to expand its use, and we went looking for schools that could benefit from it. At private schools funded by wealthier parents or philanthropic donors, principals would lead us to sparkling classrooms with rows of well-kept computers. But those weren't the schools that needed a boost. Their students would do well with or without MultiPoint. In the schools where help was most needed–where administrators were apathetic or underfunded, where teachers were absent or overloaded, or where students learned little and rarely graduated–it was impossible for MultiPoint to gain a foothold.

One visit I made to a government primary school just outside of Bangalore illustrates why. The headmaster unlocked a large metal cabinet to show me where he kept the school's personal computers. Inside, desktop PCs, monitors, and keyboards were piled shoulder high, somehow caked in dust even though they weren't out in the open. He explained that the PCs had been allocated to each school in the district two years before. The equipment had been received with excitement. The headmaster had cleared a room in his spartan cement-block building for a computer lab. Classes visited the lab one after the other, and students, crowding five or six to a PC, found games to play. The teachers, however, complained that the games didn't follow the curriculum, and in any case, they didn't know how to incorporate digital tools for teaching. Then, within weeks, the equipment began to fail. Power surges were probably to blame.[5] The school had no IT staff, and there was no budget for technical support. Soon after, the machines were locked away, and the computer lab was repurposed. The PCs were just taking up storage space, but the headmaster couldn't get rid of them. As state assets, they might be subject to inspection.

The situation wasn't unusual. Many schools had neither staff nor finances for ongoing technical support. Computer budgets in education tend to pay for hardware, software, and infrastructure, but they neglect the ongoing costs of storage, upgrades, troubleshooting, maintenance, and repair. And PCs need a lot of care in the hot, dusty, humid conditions of rural Indian schools.

Meanwhile, teachers who had PCs dumped into their classrooms felt like seafaring captains suddenly asked to pilot a jumbo jet, all while the unruly passengers are given free access to the controls. For teachers already struggling to keep their students engaged, a computer was less help, more hindrance.

In the course of five years, I oversaw at least ten different technology-for-education projects. We explored video-recorded lessons by master teachers; presentation tools that minimized prep time; learning games customizable through simple text editing; inexpensive clickers to poll and track student understanding; software to convert PowerPoint slides into discs for commonly available DVD players ; split screens to allow students to work side by side; and on and on.[6] Each time, we thought we were addressing a real problem. But while the designs varied, in the end it didn't matter–technology never made up for a lack of good teachers or good principals. Indifferent administrators didn't suddenly care more because their schools gained clever gadgets; undertrained teachers didn't improve just because they could use digital content; and school budgets didn't expand no matter how many "cost-saving" machines the schools purchased. If anything, these problems were exacerbated by the technology, which brought its own burdens.

These revelations were hard to take. I was a computer scientist, a Microsoft employee, and the head of a group that aimed to find digital solutions for the developing world. I wanted nothing more than to see innovation triumph, just as it always did in the engineering papers I was immersed in. But exactly where the need was greatest, technology seemed unable to make a difference.

Textbooks of the Air

We were hardly the first to think our inventions would transform education. Larry Cuban, a veteran inner-city teacher and an emeritus professor at Stanford, has chronicled the technology fads of the past century. As his examples show, the idea that technology can cure the ills of society is nothing new. As early as 1913, Thomas Edison believed that "the motion picture is destined to revolutionize our educational system."[7] Edison estimated that we only learned 2 percent of the material we read in books, but that we could absorb 100 percent of what we saw on film. He was certain that textbooks were becoming obsolete.

In 1932, Benjamin Darrow, founder of the Ohio School of the Air, made a similar claim for radio. He said that the medium would "bring the world to the classroom . . . [and] make universally available the services of the finest teachers, the inspiration of the greatest leaders." Radio would be "a vibrant and challenging textbook of the air."[8]

In the 1950s and 1960s, it was television. President John F. Kennedy convinced Congress to authorize $32 million for classroom television programs. For a few years, American Samoa based its entire school system on televised instruction. President Lyndon B. Johnson approved. "The world has only a fraction of the teachers it needs," he said. "Samoa has met this problem through educational television."[9]

All of these predictions sound achingly similar to today's claims for digital technology. If history is a guide, new technologies will be absorbed by schools but will do little in the end to advance education. Audiovisual teaching aids are common in modern classrooms, to be sure, but they have hardly revolutionized learning. It now seems quaint, even silly, to think that a generation hung its educational hopes on the boob tube. Television was supposed to uplift millions. Instead, millions sit in thrall to the Kardashians.

Maybe, though, digital is different. After all, real education involves two-way interaction, while broadcast media is only one-way. Don't computers, the Internet, and social media offer something that television doesn't?

Rigorous studies say no. The economist Ana Santiago and her colleagues at the Inter-American Development Bank found no educational advantage in a One Laptop Per Child program in Peru. Three months after an enthusiastic nationwide rollout, the novelty factor had worn off, and each week saw less use of the laptops. Even after fifteen months, students gained nothing in academic achievement.[10] Another team of researchers found similar results in Uruguay: "Our findings," they said, "confirm that the technology alone cannot impact learning."[11]

Economist Leigh Linden at the University of Texas at Austin conducted experimental trials in India and Colombia. He found that, on average, students exposed to computer-based instruction learned no more than control groups without computers.[12] His conclusion? While PCs can supplement good instruction, they don't substitute for time with real teachers.

One of our research partners in India was the Azim Premji Foundation, a nonprofit organization that at the time ran the world's largest program involving computers in education. In 2010, its CEO, Anurag Behar, published a brave article in an Indian affiliate of the *Wall Street Journal*. Casting doubt on his own group's work with computer labs in over 15,000 schools, he wrote, "At its best, the fascination with [information and communication technology] as a solution distracts from the real issues. At its worst, ICT is suggested as substitute to solving the real problems."[13]

We Don't Need No Digitization

The lessons of a place with circuit-frying electrical supply and no running water might seem irrelevant for those of us who live in the developed world. American schools, though, suffer similar fates with technology. In 2001 and 2002, Mark Warschauer, a professor at the University of California, Irvine, and one of the world's experts on technology in the classroom, led a study of eight schools in California that spanned rich and poor socioeconomic groups.[14] Foreshadowing my experience in India, he found that US schools also had problems maintaining technology and using it meaningfully.

Warschauer heard many teachers complain that computers in the classroom had doubled their workload. Not only did teachers have to design lesson plans involving computers, they also had to write low-tech backup plans in case of technology failures, which were frequent.

Even when it worked, technology wasn't necessarily being used well. In one class, Warschauer witnessed students typing names of countries into a search engine, clicking on whatever web pages came up, and aimlessly copying snippets of text into word-processing software. He wrote, "Although the students could be said to be performing the task of searching for material on the Web, they were not developing any of the cognitive or information literacy skills that such a task would normally involve."

Warschauer also found that poorer districts had more difficulty with the equipment. What mattered wasn't the technology–all of the schools had about the same number of computers per student and similar access to the Internet. But, he wrote, "placing computers and Internet connections in low-[income] schools, in and of itself, does little to address the serious educational challenges faced by these schools. To the extent that an emphasis on provision of equipment draws attention away from other important resources and interventions, such an emphasis can in fact be counterproductive."[15]

Other scholars, journalists, and educators have taken a hard look at electronics in the American classroom and found them wanting. In

The Dumbest Generation, Emory University professor Mark Bauerlein cites statistic after statistic showing that "digital natives"–millennial children who have never known a life without the Internet–aren't doing any better in school than their parents did. He rails against the fetish we make of technology: "It superpowers [students'] social impulses, but it blocks intellectual gains."[16] Todd Oppenheimer, in *The Flickering Mind*, grieves over his visits to computerized schools across the country. All too often, he finds digital education to be about cutting and pasting graphics into PowerPoint.[17] The school board president of the Liverpool Central district near Syracuse, New York, Mark Lawson, canceled a disappointing school laptop program after a run of seven years. There

"was literally no evidence it had any impact on student achievement–none. . . . The teachers were telling us when there's a one-to-one relationship between the student and the laptop, the box gets in the way. It's a distraction to the educational process."[18]

In other words, even in America, where infrastructure is reliable and technology is plentiful, computers don't fix struggling schools.

Nevertheless, when I returned to the United States in 2010, I came home to a country that was on a whole new kick about technologies for education. Secretary of Education Arne Duncan gave talk after talk urging more technology in the classroom. One keynote he delivered in 2012 was indistinguishable from a tech-company sales pitch. In it he mentioned technology forty-three times: "technology is the new platform for learning"; "technology is a powerful force for educational equity"; "technology-driven learning empowers students and gives them control of the content"; "technology . . . provides access to more information through a cell phone than I could find as a child in an entire library." (In the same speech, he mentioned teachers only twenty-five times.)[19]

Marc Prensky is the consultant who coined the term "digital natives." He claimed that "today's students think and process information fundamentally differently from their predecessors." Immersed in devices from birth, they are growing up in a new world that their digital-immigrant parents don't fully understand. His recommendation? We should teach digital natives in the language they were born in: "My own preference for teaching Digital Natives," he wrote, "is to invent computer games to do the job, even for the most serious content."[20]

Egged on by the chorus of support, America is in an orgy of educational technologies despite scarce evidence that they improve learning. In 2013, the Los Angeles Unified School District announced a $1 billion program to distribute iPads to all of its students.[21] Donors flock to support the online Khan Academy, where the disembodied voice of Salman Khan accompanies video-recorded blackboard instruction. And MOOCs–massively open online courses–from Harvard, MIT, Stanford, and other universities boast about the millions of people from around the world taking their free classes.

The fever is contagious. Despite everything I learned in India, I wasn't immune to it. I was once on a panel at MIT with Negroponte where I outlined my hard-won lessons about technology for education. He didn't like what I said, and he went on the offensive. But he did it with such confidence and self-assurance that, as I listened, I felt myself wanting to be persuaded: Children *are* naturally curious, aren't they? Why *wouldn't* they teach themselves on a nice, friendly laptop?

As I heard more of the technology hype, however, I realized that it didn't engage with rigorous evidence. It was empty sloganeering that collapsed under critical thinking.

Take Negroponte's claim that children are natural learners who will teach themselves with well-designed gadgets. Its subversive edge is part of its charm. Pink Floyd lyrics echo in the mind: "We don't need no education; we don't need no thought control." But even casual observation suggests that the truth is otherwise. When left alone with technology, few children open up educational apps. What they really want is to play Angry Birds. And teenagers are not that different. Los Angeles's iPad initiative hit an early glitch when older students hacked the tablets' security software and gained access to games and social media.[22]

Another highly touted project that doesn't measure up is called the Hole-in-the-Wall. Its main proponent is Sugata Mitra, a professor of educational technology at Newcastle University. He regales audiences with what he says happened when he embedded a weatherproofed

PC in the wall of a New Delhi slum. Without any supervision, local children started using the computer. They taught themselves to open applications, draw pictures, and use the Internet. In later studies, Mitra made the astonishing claim that his brand of "minimally invasive education" allowed children in poor villages to learn English and molecular biology entirely on their own.[23] Mitra went on to become an internationally celebrated speaker and won the 2013 TED Prize.

But some who visit Mitra's Hole-in-the-Wall sites find that they are unused, defunct, or occupied by older boys playing video games.[24] Payal Arora, a professor of media and communication at Erasmus University in Rotterdam in the Netherlands, found one village where instead of the reputed computers, there was only a "cemented structure in which there are three gaping holes." Several years after the computers were installed, "few of the people [in the village], including the students, had any recollection of the project." One local teacher "recalled a few boys using these kiosks, but 'usually for things like games, that's all.'" Confronted with these points, Mitra softened his position, admitting, "It is certainly incorrect to suggest that free access to outdoor-located PCs is all that is involved" in real learning.[25]

A 2013 study by Robert Fairlie and Jonathan Robinson–economists with no stake in technology–slams a heavy lid on the sarcophagus for the quixotic idea that children will teach themselves digitally. In an experimental trial involving over 1,000 students in grades 6 through 10 in America, they found that students randomly selected to receive laptops for two years certainly spent time on them, but that the time was devoted to games, social networking, and other entertainment. And whatever merit these activities might have in theory, in practice those with laptops did no better "on a host of educational outcomes, including grades, standardized test scores, credits earned, attendance, and disciplinary actions," than did a control group without computer access at home.[26] In other words, unfettered access to technology doesn't cause learning any more than does unfettered access to textbooks.

Technology advocates ignore studies like this. Instead they prey on parental fears. Secretary Duncan insists that "technological competency is a requirement for entry into the global economy," hinting that our children will be at a disadvantage if they grow up without computers.[27]

But do students need to be steeped in technology to be competitive? Duncan is himself proof that it's not. By his own admission, he grew up in a "technology-challenged household." Apparently, when he was young, his family didn't even own a television, to say nothing of a PC.[28] Like most of today's leaders above the age of forty-five, Duncan wasn't exposed to digital technology when he was young, yet we can be sure that his mother is proud of his accomplishments in the twenty-first century.

Years of data from the world's most credible educational yardstick show that technology in the classroom has little to do with good scores. The Program for International Student Assessment (PISA) is the Olympics of formal education. Participating countries administer standardized academic achievement tests to their fifteen-year-olds, allowing cross-country comparisons in several subjects. South Korea happens to be both high-tech and high-performing, but Finland and China consistently outperform other countries despite low-tech approaches.[29] In a 2010 report, PISA analysts wrote that "the bottom line is that the quality of a school system cannot exceed the quality of its teachers," regardless of available educational resources such as computers.[30]

Anyone can learn to tweet. But forming and articulating a cogent argument in any medium requires thinking, writing, and communication skills.[31] While those skills are increasingly expressed through text messaging, PowerPoint, and email, they are not taught by them. Similarly, it's easy to learn to "use" a computer, but the underlying math skills necessary for accounting or engineering require solid preparation that only comes by doing problem sets–readily accomplished with or without a computer. In other words, there's a big difference between learning the digital tools of modern life (easy to pick up and getting easier by the day, thanks to improving technology) and learning the critical thinking skills necessary for an information age (hard to learn and therefore demanding good adult guidance). If anything, it's less useful to master the tools of today, because we know there will be different tools tomorrow.

What Wise Parents Know

For about a month in the spring of 2013, I spent my mornings at Lakeside School, a private school in Seattle whose students are the scions of the Pacific Northwest elite. The beautiful red-brick campus looks like an Ivy League college and costs almost as much to attend. The school boasts Bill Gates among its alumni, and there is no dearth of technology. Teachers post assignments on the school's intranet; classes communicate by email; and every student carries a laptop (required) and a smartphone (not).

In this context, what do parents do when they think their children need an extra boost? I was there as a substitute tutor for a friend whose students spanned the academic spectrum. A few of them were taking honors calculus. They were diligent but wanted a sounding board as they worked on tough problems. Others, sometimes weighed down by intensive extracurricular activities, struggled in geometry and algebra. I would review material with them and offer pointers as they did assignments. Yet another group required no substantive help at all. They just needed some prodding to finish their homework on time. Despite their differences, the students had one thing in common: What their parents were paying for was adult supervision.

All of the content I tutored is available on math websites and in free Khan Academy videos, and every student had round-the-clock Internet access. But even with all that technology, and even at a school with a luxurious 9:1 student-teacher ratio, what their parents wanted for their kids was extra adult guidance. If this is the case for Lakeside students with their many life advantages, imagine how much more it must be the case for the world's less privileged children.

If the Labors of Hercules had an intellectual equivalent, it would be modern education. By the end of high school, we expect a student to learn about 60,000 words; read *To Kill a Mockingbird*; learn the Pythagorean theorem; absorb a national history; and have peered through a microscope. Advanced students will put on Greek tragedies; rediscover the principles of calculus; memorize the Gettysburg Address; and measure the pull of gravity. In effect, students have twelve years to reconstruct the world's profoundest thoughts–discoveries that history's greatest thinkers took centuries to hit upon.

This is not casual play, and it requires directed motivation. It doesn't matter what flashy interactive graphics exist to teach this material unless a child does the hard internal work to digest it. To persevere, children need guidance and encouragement for all the hours of a school day, at least nine months of the year, sustained over twelve years. Electronic technology is simply not up to the task. What's worse, it distracts students from the necessary effort with blingy rewards and cognitive candy. The essence of quality children's education continues to be caring, knowledgeable, adult attention.

Miracle or Mirage?

Yet, it can't be true that technology never helps education. That doesn't square with what my team found with the MultiPoint pilot, or, for that matter, with any project where formal trials prove a technology's value. There is plenty of reliable research in which students with technology gain something over those without.

Indeed, Negroponte is persuasive because he speaks with deep conviction. He's a true believer. Negroponte's backstory involves a rural Cambodian village where out of a charitable impulse he handed out laptops to twenty children. When he found both the students and their families making innovative use of them, One Laptop Per Child was born.[32]

And Professor Warschauer at UC Irvine, no utopian when it comes to technology, found that in some American schools with one-to-one laptop programs, the students became better writers. They wrote more, revised more, and got more frequent feedback from teachers.[33]

And what about the evidence under my nose? The proof of technology's value in my own learning as a research scientist? It is thanks to the Internet that I can look up papers without a trip to the library. It is thanks to email that I can stay in touch with colleagues on the other side of the world. And it is thanks to Wikipedia that I can brush up on knowledge long forgotten or never learned.

So in some cases, technology does help–but not with the consistency required to fix larger social problems. The MultiPoint experience was repeated in all of my team's other projects–in agriculture, in health care, in governance, in entrepreneurship. On the one hand, it was easy to develop innovations that had some merit; on the other hand, those same innovations rarely led to large-scale benefits. How could this be? It was a paradox. I was missing an explanation of how and why machines contribute–or don't–to social change.

Actually, modern society as a whole lacks a good framework for thinking about technology's social impact. As children, we learn how our bodies work in biology classes and how our government works through civics lessons. Computer courses, though, only teach us how we can use the devices, not how the devices affect us. As adults, we're inundated with news about Facebook revolutions in the Arab Spring, long queues for the latest iPhone, and email spying by the National Security Agency. Yet we have no consensus view of the technologies' net effect.

Toward the end of my five years in India, I had a glimpse of a hypothesis. I knew there was a way to make sense of the apparent contradiction whereby isolated successes weren't easy to replicate elsewhere . But since I worked at a company whose soul was software, I kept wanting to see the technology prevail. I felt disloyal doubting its value. As Upton Sinclair said, "it is difficult to get a man to understand something, when his salary depends upon his not understanding it."[34] I needed some distance, and I needed some time. So in early 2010 I left Microsoft to join the School of Information at Berkeley. The dean, AnnaLee Saxenian, had arranged a research fellowship for me. At her school, people not only built technologies but also studied how they affected society. Technology's impact was complex, but I hoped to find a concise way to understand when it was good, when it was bad, and when we could know in advance.

CHAPTER 2

The Law of Amplification

A Simple But Powerful Theory of Technology's Social Impact

Nakkalbande is a small slum community in the southern part of Bangalore. Hidden within the upper-middle-class neighborhood of Jayanagar, it's formed around a single, straight alley that is covered by a canopy of grand old trees that have survived the city's aggressive road construction. The unpaved alley is strewn with plastic debris and the occasional dead rat. As slums go, though, it's doing all right. Instead of the improvised tarp-and-tree-branch shelter you might see elsewhere, most of the houses in Nakkalbande are one-or two-room cinder-block structures. Residents have lived there for decades.

Nakkalbande is where I spent my Saturdays soon after moving to India in late 2004. I volunteered for a nonprofit called Stree Jagruti Samiti, the Society for Women's Empowerment. Its leader was a middle-aged matriarch named Geeta Menon, who had a mischievous chuckle and a gleam in her eye that wouldn't be brought down by the tired droop in her shoulders. For over fifteen years, she had worked as an activist, organizing the women and girls of several slum communities. She was known to storm into police stations with groups of women. They would demand that the officers take action against, say, a corrupt rations dealer. (Rations shops in India are licensed to sell subsidized food and kerosene to households below the poverty line, but they often profit from sale of inventory to other retailers.)

At Menon's suggestion, I taught a computer literacy class for girls. I didn't speak any of the languages they spoke–Hindi, Kannada, or Tamil–so I recruited a college student to translate and assist me. On the first day, eight or nine teenage girls dressed in pastel *salwar kameez* gathered into a small, windowless building that was reserved for community activities. I brought a laptop and set it up under a framed picture of a blue-skinned Krishna playing his flute.

When my assistant and I told the girls they were going to learn how to use a computer, their eyes widened, and a collective shriek filled the room. Over several weeks, we showed them the basics of word processing, PowerPoint, spreadsheets, and other software. At first they gawked at simple things such as moving the cursor, using the touchpad, and clicking to cause action on the screen. Novelty, though, quickly gave way to familiarity. Soon they were fighting over who would get to draw next using a painting application. Like computer novices everywhere, they took delight in converting words into every conceivable color and font. Their enthusiasm was infectious, and I looked forward to the classes.

By the third or fourth session, though, we hit a wall in what they could learn. Everyone was able to type her name in English as well as in Kannada, but the girls weren't interested in writing anything beyond that. PowerPoint became known as the software that allowed them to create fancy 3D text. And spreadsheets thoroughly bored everyone except for two girls who sensed something extraordinary in self-computing arrays of numbers. We contrived activities to both entertain and educate, but, in practice, it was hard to go beyond entertainment.

I began to understand why this was the case as I learned more about the girls' personal lives. Their days were crammed with school and chores. They worked part-time as servants in middle-class homes. With so many adult responsibilities, they saw the computer class as a break from a life of constraints. Some would linger afterward to teach me folk games as a way to extend their freedom. No one mentioned any serious hobbies, though, and the one thought of their future was about who would be arranged for them as husbands. Despite Menon's best efforts, fourteen or fifteen wasn't an unusual age for marriage. The girls expected to become housewives in short order, and few would continue school beyond eighth or ninth grade.

Originally, Menon and I had vague hopes that the computer classes would help the girls gain access to work other than as household servants. But even for entry-level positions, employers wanted a solid education first, white-collar soft skills second, and then only on top of that, computer literacy. With just one class per week–their parents didn't allow more–we couldn't have taught them more employable skills such as programming or data entry.

At the end of the course, we took the girls to visit a local Internet café, but little of lasting value came of the trip. Like many such spots in urban India, this was a dingy place with two or three old desktop computers running outdated versions of Windows. (Even as of 2013, Windows 98 was a common sight in Indian Internet cafés.) For about 10 rupees (roughly 20 cents), you can use an Internet-connected PC for an hour, but you get what you pay for. It can take half a minute, for example, to load the bare-bones Google home page. In formal studies later on, Nimmi Rangaswamy, a member of my research team, found that Internet café clientele is dominated by young men chatting, playing video games, and consuming pornography; many owners install private booths for the purpose.[1] As a result, Indian families think of cybercafés as sleazy places. Women and girls aren't encouraged to visit them.

Still, the exposure to computers did have some unexpected effects. The two girls who found spreadsheets fascinating vowed to stay in school for as long as they could, in spite of parental pressures to take on more chores. They recognized that they needed to know more in order to take advantage of the technology. But then, another girl dropped out of the class within a few weeks. She told me her parents didn't want her to learn too much because that would raise her dowry. Families with sons expect dowries as something like a down payment for the costs of keeping a wife. The fear is that a more educated bride will have higher expectations and require more upkeep. (Apart from its patriarchal conception, this traditional calculus doesn't account for the possibility that an educated wife could bring in her own income, as happens more and more across India.)

I didn't think of the computer course as a formal research project, so I didn't keep detailed track of the outcomes. When I look back, though, I realize that the class foreshadowed what I'd soon find in my own research: the initial optimism that surrounds technology, the doubt as reality hits, the complexity of outcomes, and the unavoidable role of social forces.

The Ferocious Field of Technology and Society

Technology is powerful, but in India it became clear to me that throwing gadgets at social problems isn't effective. When I came back to the United States, I sought to understand why.

As a computer scientist, my education included a lot of math and technology but little of the history or philosophy of my own field. This is a great flaw of most science and engineering curricula. We're obsessed with what works today, and what might be tomorrow, but we learn little about what came before.

So at the University of California, Berkeley, I met with dozens of professors who had studied different aspects of technology and society. I spent hours tracking down dusty, bound volumes in the stacks of libraries across campus. And here is what I learned.

Theorists, despite many fine shades of distinction, fall roughly into four camps: technological utopians, technological skeptics, contextualists, and social determinists. These terms will be defined in a moment, but one thing that jumped out was that the scholars fought like Furies. For example, the economic historian Robert Heilbroner wrote, "That machines make history in some sense . . . is of course obvious."[2] This view is called technological determinism, because it implies that technology determines social outcomes. But if some find it obvious, it is nevertheless ridiculed by critics. Philosopher Andrew Feenberg responded with sarcastic sympathy, writing that "the implications of determinism appear so obvious that it's surprising to discover that [its premises do not] withstand close scrutiny."[3]

Yet for all the debate, there is plenty of agreement, too. Utopians accept that there can be negative consequences of technology, and skeptics concede its benefits. What separates the four camps most is not facts but temperamental differences.

How to Spot a Utopian

In the *Star Trek* future, technological advances have liberated Earth from war, famine, illness, and conflict, at least among human beings. Thanks to matter replicators and dilithium crystals, food and energy are free. With nothing to fight over, peace and egalitarianism reign. (That's why the series needs an ample supply of aliens as plot devices.) As Captain Jean-Luc Picard explains in the movie *First Contact*, "the acquisition of wealth is no longer the driving force in our lives."[4] That is to say, in a few more centuries, advanced technology makes economics itself obsolete. Instead, people are free to focus on greater ends: "We work to better ourselves and the rest of humanity."

Star Trek is fiction, but its technological utopianism is very real. MIT Media Lab founder Nicholas Negroponte clearly shares it. So does Google chairman Eric Schmidt. In *The New Digital Age*, he and coauthor Jared Cohen wrote, "The best thing anyone can do to improve the quality of life around the world is to drive connectivity and technological opportunity."[5] And then there are technology cheerleaders like Clay Shirky, who shakes pom-poms for Team Digital in a book subtitled *How Technology Makes Consumers into Collaborators*.[6] Many engineers and computer scientists also hold this view. A generation ago, when young people said they wanted to "change the world" or "make an impact," they joined the Peace Corps. Now they move to Silicon Valley. They envision laying a foundation for Captain Picard's greedless future.

Utopians believe that technology is inherently a positive force, that technology shapes civilization, and that more of it is a good thing. And they have what seems like irrefutable evidence. Thanks to advances such as modern medicine, air conditioning, cheap transport, and real-time communication, middle-class people today enjoy a quality of life that kings and queens didn't have a century ago. There's a reason, utopians argue, why historical epochs are named after technologies–the Bronze Age, the Iron Age, the Industrial Age, the Information Age–and why human culture flourished after the invention of the printing press.

But whatever they say and write, what most unites utopians is how they feel about technology. They love it, and they want more. Many believe that *every* kind of problem can be solved by some invention, often one that is right around the corner. Whether the issue is poverty, bad governance, or climate change, they say things like, "[There] is no limit to human ingenuity," and "When seen through the lens of technology, few resources are truly scarce."[7] Besotted with gadgets, technological utopians scoff at social institutions like governments, civil society, and traditional firms, which they pity as slow, costly, behind the times, or all of the above.

I sympathize with the utopians because I was one myself. When I started the computer class in Nakkalbande, it was in the hopes that exposure to the technology would improve lives. And my research looked for ways to use technology to alleviate poverty.

A Curmudgeonly Skepticism

But time after time, I realized that technology alone never did the trick. Whether it was MultiPoint in India or laptops in America, inventing and spreading new devices didn't necessarily cause social progress.

Technology skeptics would harrumph and point out that aspects of the *Star Trek* future are already with us. Thanks to agricultural technologies, America produces more than enough food to feed everyone in the country, and the food is cheap. Yet, almost 5 million children in the United States suffer from food insecurity in any given year.[8] Indeed, there is enough food to feed the whole world, but hunger persists. About one in eight people is malnourished; that's 840 million people eating less than they need.[9] Evidently, technological plenty doesn't mean plenty for everyone.

Skeptics believe that technology is overhyped and often destructive. Nicholas Carr, author of *The Shallows*, suggests that the fast-twitch, hyperlinked Internet not only erodes our ability to think deeply, but also traps us like a Siren: "We may be wary of what our devices are doing to us, but we're using them more than ever." His book is ominously subtitled *What the Internet Is Doing to Our Brains*. In *The Net Delusion: The Dark Side of Internet Freedom*, Evgeny Morozov catalogs the myriad ways in which the Internet boosts, rather than contains, the power of repressive regimes: in China, social media is a tool for disseminating Communist Party propaganda; in Azerbaijan, webcams installed at election stations frightened citizens into voting for state-sponsored incumbents.[10] In Iran, the chief of national police acknowledged a chilling fact of their anti-protest efforts: "The new technologies allow us to identify conspirators."[11]

Technology skeptics like to point out unintended consequences. Jacques Ellul, for example, warned of the dangers of information overload back in 1965. "It is a fact that excessive data do not enlighten the reader or the listener," he wrote. "They drown him."[12] Neil Postman suggested that broadcast media have created a culture that is "amusing itself to death,"[13] like mythological lotus-eaters, or the soma-sedated characters of Aldous Huxley's *Brave New World*. And Harvard professor Sheila Jasanoff has voiced the concerns of many in calling out climate change as a by-product of fossil-fuel-driven technologies.[14] Incidentally, digital technologies play a shockingly large part in carbon emissions. One study estimated that in 2007, electronics accounted for 3 percent of carbon emissions globally and 7.2 percent of all electricity usage.[15] In the United States in 2013, the data centers that store and distribute online content accounted, on their own, for about 2 percent of total electricity use.[16] All of these figures are projected to grow.[17]

If skeptics are pessimistic, though, many of them share the utopians' belief that technologies embody moral and political values. But where utopians see the promise of greater freedom and prosperity, skeptics see weakness, folly, and corruption. The economic efficiency of factories and assembly lines leads to a dehumanized society. High-tech entertainment prompts us to judge everything by its marketability. Social media turns us into zombies of "continual partial attention."[18]

As for practical action, skeptics are less united than their utopian counterparts. They span a spectrum from neo-Luddites who would destroy technology to those who can't quite give up their smartphones. At one extreme is author and activist Derrick Jensen, who wrote, "Every morning when I wake up I ask myself whether I should write or blow up a dam."[19] Carr invoked a poet's call for resistance, hoping that "we won't go gently into the future our computer engineers and software programmers are scripting for us."[20] And some just throw up their hands. Ellul could see no easy solution: "It is not a matter of getting rid of it, but, by an act of freedom, of transcending it. How is this to be done? I do not yet know."[21]

Not Good, Not Bad, Not Neutral

Utopians and skeptics have catchy rhetoric, but most reasonable people can see that the truth is neither *Star Trek* nor *Brave New World*. It's probably a mixture of both. Melvin Kranzberg, a historian of technology, embraced technology's apparent contradictions. "Technology," he wrote in 1986, "is neither good nor bad; nor is it neutral."[22] This enigmatic statement captures what is probably the most common view among scholars of technology today: Its outcomes are context-dependent. Technology has both positive and negative impacts because technology and people interact in complex ways.

But contextualist explanations are also unedifying. To stop at context dependency is to say very little altogether. The lessons tend to follow the lines of "more research is needed"; "it's case by case"; or "it's nuanced"–ivory-tower code for "it's so complicated, there couldn't possibly be any worthwhile generalizations." As a proponent of one contextualist theory claimed, "explanation does not follow from description."[23]

The Human Factor

Utopians, skeptics, and contextualists are each right in limited ways. The fifty-odd technology projects I oversaw in India produced a range of outcomes. A few improved people's lives. The utopians would have cheered. A few wasted time and resources. The skeptics would have said, "I told you so." The majority fell into a middle ground where they succeeded as research projects, but benefits beyond that were limited. The contextualists would have nodded in sympathy.

But was there some other way to interpret these outcomes? As I looked for some structure to our findings, three factors emerged as necessary for real impact.

The first is the dedication of the researcher, not to research outcomes but to concrete social impact. Of all the projects I oversaw, the one that continues to affect the most lives is called Digital Green. It uses how-to videos featuring local farmers as a teaching aid to help instruct other farmers about better agriculture. Today the Indian Ministry of Rural Development is taking Digital Green to 10,000 villages, and the Ethiopian government has begun experimenting with it as well. None of this would be happening without Rikin Gandhi, who led the project. Gandhi has many talents, but what stands out is his single-minded focus on supporting smallholder farmers. Instead of designing the electronic version of a Rube Goldberg machine–which is what feature-happy technologists tend to do–he stuck with simple, off-the-shelf devices. Then, after we established Digital Green's effectiveness, he left his research job to start a nonprofit organization. Without Gandhi's devotion to social impact, Digital Green wouldn't be much more than a research paper.

The second factor is the commitment and capacity of the partner organization. In my research group, we looked for capable, well-intentioned partners who had rapport with the communities we wanted to work with. Sometimes, though, we'd misjudge an organization and find ourselves stymied by its dysfunctions. In one project, we partnered with a sugarcane cooperative in a rural district three hours away from Bombay. We upgraded its communication infrastructure by replacing a creaky network of old personal computers with low-cost mobile phones. The new system worked, and farmers loved it. Had the cooperative rolled it out to all of its villages, it would have saved them tens of thousands of dollars every year.[24] Yet an internal rivalry kept us from expanding beyond the pilot. (And as researchers, we lacked the patience and charm to iron out the discord.) The technology worked perfectly, but institutional politics hampered deployment. Good partners were important, even with good technology.

The third factor lies with intended beneficiaries. They must have the desire and the ability to take advantage of the technology provided. Sometimes they don't. In India, we worked with poor people who lacked basic health care and hygiene, so we thought it would be useful to offer the right information at the right time. But would-be beneficiaries hesitated to follow even the simplest advice. Women wouldn't take iron pills because of the bitter taste. Households wouldn't boil water because of the extra effort. Fathers would lose infants to minor illnesses because they balked at hospital charges of as little as 50 rupees (about $1, potentially a day's wages). In other words, they were like any of us who fail to exercise and eat well despite knowing that we should. It didn't matter whether we delivered the information via text messages, automated voice calls, entertaining videos, or interactive apps. Technology by itself didn't budge social and psychological inertia.

These factors suggest that the contextualists are right. Context definitely matters. All three factors, though, point to *human* context as what matters most. Or, to put it another way, the technology isn't the deciding factor even in a technology project. Of course, good design trumps poor design, but beyond some level of functionality, technical design matters much less than the human elements.[25] The right people can work around a bad technology, but the wrong people will mess up even a good one.

This is consistent with a fourth camp of technology-and-society scholarship sometimes called social determinism.[26] Versions of it are known as "the social construction of technology" and the "instrumental view" of technology. These and related theories emphasize that technology is molded and wielded by people. People decide the form of technologies, the purposes of their use, and the outcomes they generate. Social determinism rests on the plain fact that it is people who act and make decisions–technologies do not.[27]

But if social determinism is commonsense, it's not quite enough. It says little about how much change follows in the wake of invention. So while I felt close kinship with social determinists, something was still missing.

It's All Geek to Me

If you've ever landed on a webpage in a language you can't read, you have an idea of what it means to be illiterate in a digital world. You can see that there's a whole universe bursting with possibility, but none of it makes sense. You might recognize a few photos here and there, but your curiosity is piqued only to bang into a wall of indecipherable gibberish.

That was the experience of those we worked with who couldn't read. It was true of the mothers of the students I taught in Nakkalbande, some of whom would pop into an occasional class to see what their children were up to. So one agenda for our research was digital interfaces for nonliterate users. In 2005 I hired Indrani Medhi, a designer who threw herself into the research and emerged within a few years as the world's expert on what we called "text-free user interfaces."

Medhi conducted much of her research in Nakkalbande. She got along well with Menon and shared her combination of toughness and empathy. Medhi was quick to befriend her research subjects–mostly women from poor families who earned $20 to $40 a month doing informal household work. Through them, Medhi found that illiteracy didn't always mean innumeracy, at least in those communities. Many of the women could read numbers, even if they sometimes confused "2" and "5." With a colleague, Archana Prasad, she also found that respondents understood cartoon drawings best, finding them less confusing than either simplified icons or photographs.[28] These and other discoveries fed directly into Medhi's designs.

Medhi and I had frequent discussions about her work, and some themes came up repeatedly. One was that illiteracy wasn't black and white–it was a spectrum. Some people couldn't read at all, others knew the alphabet, and still others could sound out words but couldn't read a newspaper. Another point was that users differed considerably in their responses to the same interface. Some people zipped through Medhi's text-free interfaces and even seemed to enjoy the process. Others were hesitant and slow and required encouragement to continue.

These traits seemed correlated: More literate people were more adept with computer interfaces, even when the interfaces contained no text. To investigate further, we ran a study in which participants were first given tests of literacy and abstract reasoning and then asked to perform a simple task on a computer.[29] The task was to navigate a menu interface, which we knew would be a challenge. The respondents were asked to find specific household items among cartoon graphics organized in one of two ways. In the first, the objects were laid out so as to be visible all at once, but in a random order. In the second, they were organized as a series of nested items, similar to files put into folders on a computer. In the nested interface, bangles, for example, could be found first by clicking a graphic indicating things you wear (versus things you use), and then jewelry (versus clothing), and then hands (versus face or feet).

The research validated our hunches. First, the degree of literacy correlated with the measure of abstract reasoning capacity. Second, all of the participants were quicker to find items in the single unorganized list than in the nested hierarchies. And third, on both navigation tasks–flat list and nested hierarchies–those who scored higher on the tests of literacy and reasoning outperformed their lower-scoring peers.

So whatever level of intelligence and education a person already had correlated with their facility with simple computer tasks. People with greater education and cognitive capacity were better able to use the technology. It would be careless to generalize too much from this one finding, but over the years, I saw many similar results. In a related study, supplying textual hints along with audio and graphics helped the literate more than the semiliterate and the semiliterate more than the nonliterate. Another group examined mobile phones and Indian women micro-entrepreneurs. The researchers found that the most ambitious and self-confident women benefited most from mobile phones. And a study of Tanzanian health-care workers showed that their visits to patients increased with text-message reminders, but only if they were also overseen by human supervisors.[30]

In other words, what people get out of technology depends on what they can do and want to do even without technology. In retrospect this seems self-evident, but it wasn't a major theme in technology and society literature.[31]

The Eureka Moment

So theories of social determinism say that technology is put to use according to underlying human intentions. At the same time, the degree to which technology makes an impact depends on existing human capacities. Put these ideas together and *technology's primary effect is to amplify human forces.*[32] Like a lever, technology amplifies people's capacities in the direction of their intentions. A computer allows its user to perform desired knowledge tasks in a way that is faster, easier, or more powerful than the user could without technology. But how much faster, more easily, and more powerfully is in some proportion to the user's capacity. A mobile phone allows people to perform desired communication tasks across greater distances, with more people, and at greater frequency than would be possible without one. But whom one can communicate with and what one can expect of them depends on one's existing social capacity.

The idea is so simple and so widely applicable that I have come to think of it as technology's Law of Amplification. It was at work among the girls in Nakkalbande. Most had little conscious intention to learn or improve knowledge skills, and social forces such as the expectation to marry early impeded their interest. As a result, there was little productive force for the technology to amplify. But the two older girls who recognized the value of education had some inner flame fanned by the laptop. I could imagine that, with luck and persistence, they might have a chance at a different life.

Amplification also resolved some of the apparent paradoxes of my research. Why did MultiPoint, for example, work in our pilots, but not when we took it to other schools? It was because our positive results relied on special conditions that we had imposed. For our trials, we had deliberately chosen partner schools with capable teachers and principals. As a result, the students were focused on learning. They followed instructions without too much distraction. Another critical factor was our own presence as researchers. We set up the technology ourselves. And where we found teaching capacity wanting, we filled in. In other words, we had lined up all of the social conditions favorably so that the technology had a chance to work. And on that firm base, MultiPoint increased the number of students learning from computers.

But, for expansion, we targeted subpar schools. They needed the most help, after all. We also reduced our personal involvement, since the schools would eventually have to operate without us. In the absence of good teaching and IT support, however, the technology didn't do much.

In the worst cases, technology was detrimental. More times than I'd like to admit, I'd visit a class involved in one of our projects, and something would go wrong with the technology. With no IT staff, the teacher would fumble to figure things out. The children would grow distracted. Sometimes I'd jump in to help. By the time the power was back on, the PCs rebooted, and the children once more settled into their seats, half of a fifty-minute class period was lost. It would have been better if they had stuck to pencil and paper.

In these cases, we see vividly that technologies don't have fixed additive effects. They magnify existing social forces, which themselves can be good, bad, or neutral. Thus, technological utopians and skeptics are both partially right and partially wrong. Of course, this means it's the contextualists and the social determinists who are closest to the truth. But the Law of Amplification says something more specific and therefore more useful.

For example, amplification offers clues as to why large-scale studies of educational technology rarely show positive results. In any representative set of schools, some are doing well and others poorly. Introducing computers may result in benefit for some, but it distracts the weaker schools from their core mission. On average, the outcome is a wash.

An even bigger problem is that administrators rarely allocate enough resources to adapt curricula or train teachers.[33] Where teachers don't know how to incorporate digital tools appropriately, there is little capacity for the technology to amplify.

If a private company is failing to make a profit, no one expects that state-of-the-art data centers, better productivity software, and new laptops for all of the employees will turn things around. Yet, that is exactly the logic of so many attempts to fix schools with technology.

And what about computers outside of school? What happens when children are left to learn on their own with digital gadgets, as so many tech evangelists insist we should do? Here technology amplifies the children's propensities. To be sure, children have a natural desire to learn and play and grow. But they also have a natural desire to distract themselves in less productive ways. Digital technology amplifies both of these appetites. The balance between them differs from child to child, but on the whole, distraction seems to win out when there's no adult guidance. This is exactly what Robert Fairlie and Jonathan Robinson's 2013 study of laptops in the home shows: If you provide an all-purpose technology that can be used for learning and entertainment, children choose entertainment.[34] Technology by itself doesn't undo that inclination–it amplifies it.

Amplifying Power

Back in Bangalore, I once hosted a political science professor, whom I'll call Padma. She was interested in technology and governance. Padma was in the city to study a program that made the municipal government's finances transparent to the public. A nonprofit group had convinced the government to set up a tool that let anyone with Internet access see how city money was spent. Citizens were able to see, for example, that 5,000 rupees (~$100) was spent repairing a pothole (expensive, but not unreasonable) or that 500,000 rupees ($10,000) was spent cutting down a tree (not likely; a sign of kickbacks). The nonprofit would complain to the government about egregious spending it discovered. Sometimes it organized citizen protests. Padma had a hypothesis that technology promoted transparency and accountability, and here was a system that seemed to prove it.

When I asked her how the project turned out, though, she said the government had shut it down within a few months. Officials didn't want their graft schemes open to public inspection.

If a computer system for government transparency was taken down by the very bureaucrats it was meant to monitor, then what accountability did the technology really bring? The project showed exactly the opposite of Padma's claim. Instead of technology trumping politics, politics trumped technology. At first the technology amplified the nonprofit's activism, but the organization's power to affect government was overcome by the crooked bureaucracy's greater power to turn off the technology.

Looking back at experiences like this, I saw that the Law of Amplification explained much more than just the fate of technology in education. It applies to a host of other situations. In 2011, earthshaking world events provided a unique testing ground.

Facebook Devolution

In what is now a well-worn story, Wael Ghonim, a thirty-year-old Google executive, used Facebook to help organize the protests that toppled Hosni Mubarak in Egypt. Today, it's hard to speak of the Arab Spring without calling to mind the phrase "Facebook revolution."

In early 2011, Facebook had roughly 600 million users. Almost 10 percent of the world population was on it.[35] Speculations about its initial public offering swirled, and *The Social Network*, a movie telling one version of its origins, was in theaters. With this buzz ringing in their ears, journalists and bloggers were agog over Facebook's role in Egypt. A day before the January 25 protests, *Time* asked, "Is Egypt about to have a Facebook Revolution?" It cited the 85,000 people who had pledged on Facebook that they would march.[36] Days after the first protest in Tahrir Square, Roger Cohen wrote in the *New York Times*, "The Facebook-armed youth of Tunisia and Egypt rise to demonstrate the liberating power of social media."[37] One Egyptian newspaper reported that a man named his firstborn daughter Facebook.[38]

On February 11, 2011–the day the regime folded–Ghonim told a CNN interviewer, "I want to meet Mark Zuckerberg one day and thank him. . . . This revolution started on Facebook . . . in June 2010 when hundreds of thousands of Egyptians started collaborating content. We would post a video on Facebook that would be shared by 60,000 people on their walls within a few hours. I've always said that if you want to liberate a society, just give them the Internet."[39]

If you want to liberate a society, just give them the Internet. This is a classic statement of technological utopianism. As in *Star Trek*, where technology eradicates hunger, Ghonim is saying that the Internet eradicates autocracy. Coming from someone directly involved in the revolution, it seems impolite to refute. But just as with the hype around technology for education, the case for social media as an important cause of democratic change vanishes under critical inspection.

First, let's accept that in Egypt, and earlier in Tunisia, social media contributed to the overthrow of dictators. We'll come back to exactly what that contribution was, but there's no doubt that YouTube videos and Facebook posts played a part.

But in other Middle Eastern countries, events unfolded differently. Consider Libya, for example. On February 18, 2011, just days after the rebellion started, Muammar Gaddafi dimmed communication networks in his country.[40] Maybe he had heard about the Facebook revolution next door and didn't want one in his own country. He disabled most of the Internet in Libya and did the same for phone services, mobile and landline.[41] The rebels, though, managed to coordinate nevertheless. Far from ceasing their activities, they kept fighting. Soon after, they overwhelmed Gaddafi's forces, tracked him down, and executed him in the streets.

In Syria, President Bashar al-Assad took a cue from Gaddafi. When protests began, he shut down the Internet nationwide and selectively disabled phone networks to hinder rebel communications.[42] Protests, though, continued, leading to an all-out civil war, with the rebels showing no signs of quitting even three years later. Media portrayal of Syria has long since stopped mentioning Facebook, Twitter, or YouTube.

Meanwhile, in Bahrain and Saudi Arabia, something very different happened. A few public protests were quashed in Bahrain, and the Western press hardly noticed the feeble activism in Saudi Arabia. Importantly, it wasn't for a lack of social media organization. Encouraged by the Tunisian and Egyptian revolutions, a spate of petitions and videos circulated in Saudi Arabia via Facebook and Twitter. They called for an end to absolute monarchy. But these online actions were squelched by offline forces, as reported by Madawi Al-Rasheed, a specialist in Islamist movements and Middle Eastern civil society.[43] One young activist, Muhammad al-Wadani, uploaded a YouTube video urging democracy. He was promptly arrested. Two online petitions demanding constitutional monarchy received thousands of signatures. They were ignored. A group calling itself the National Coalition and Free Youth Movement attempted to organize online. It ended up in a game of virtual Whac-A-Mole as regime security took down its websites one after the other. Protests planned on social media led nowhere.

The goal of these and other Web-based appeals was a physical protest, a "Day of Rage" on March 11, 2011. But, as Al-Rasheed wrote, "things were quiet" on the day itself: "Security forces spread through every corner and street. An unannounced curfew loomed over Riyadh and Jeddah." No protests worthy of international attention materialized.

Al-Rasheed argued that the Saudi monarchy has starved civil society in the kingdom for decades. There are no trade unions, no political parties, no youth associations, and no women's organizations. Demonstrations themselves are forbidden outright. As a result, grassroots organizational capacity is stunted. This is in direct contrast to Egypt, for example, where trade unions, nongovernmental organizations, and the Muslim Brotherhood all simmered as potent political forces despite Mubarak's oppression.[44]

The absence of protest is a non-event, so it goes unreported by mainstream news organizations. But an accurate understanding of social media's role in revolution must account for the stillborn protests of Bahrain and Saudi Arabia as much as for the successful uprisings in Tunisia and Egypt.

Did America Have a Lantern Revolution?

Combining the lessons of Tunisia, Egypt, Libya, Syria, Bahrain, and Saudi Arabia, we come to an undeniable conclusion: Social media is neither necessary nor sufficient for revolution. Claims of social media revolutions commit the classic conflation of correlation and cause. To say that the Arab Spring was a Facebook revolution is like calling the events of 1775 in America a lantern revolution thanks to Paul Revere: "One, if by land, and two, if by sea."

Actually, the tale of Revere's lanterns is itself the stuff of myth. In reality, the lantern signal was just a backup plan in case Revere was arrested and unable to sound the alarm.[45] What this story actually illustrates is that revolutionaries are contingency planners who exploit every tool at their disposal. In a telling interview about his involvement in Egypt, Ghonim noted, "They shut down Facebook. But, I had a backup plan. I used Google Groups to send a mass-mail campaign."[46] Presumably, had email also been blocked, he'd have resorted to phone calls, paper notes, and word of mouth–using the same communication tools as the 80 percent of Egyptians who have never been online. Of course, for Ghonim, "technology played a great role"–it would have been much harder for him to organize door to door. But it seems unlikely that the absence of Facebook would have prevented his activism altogether, or could have kept the rest of the country mute. Taking the broader view, Facebook was a tool of convenience for angry activists spreading the word by every available channel.

A few people tried to debunk social media's revolutionary powers. For example, just as Mubarak's regime was crumbling, Morozov's book *The Net Delusion* was released.[47] Though Morozov couldn't have known of Egypt's fate while he was writing the book, he provided some of the most insightful commentary on technology's role in Middle Eastern uprisings. His first chapter mocks the breathless hype around a supposed Twitter revolution in 2009 Iran–hype that led Hilary Clinton's State

Department to ask Twitter to postpone routine maintenance during the height of protests. (Twitter complied, and Clay Shirky wrote, "This is it. The big one. This is the first revolution that has been catapulted onto a global stage and transformed by social media."[48]) Morozov, however, cites low numbers of actual Twitter users in Iran at the time (perhaps all of sixty) and Iranian denials that Twitter had much of a role in organizing protests. He argued that Twitter was less an effective tool of protest and more a way for the outside world to eavesdrop on the events. For him, the social media narrative recalled Cold War ideas that capitalist technology would triumph over communist inefficiency, as if people in the Middle East couldn't have rebelled on their own without the gifts of American entrepreneurs. In the end, whatever was tweeted, there was no Twitter revolution in Iran.

Also among the skeptics was Malcolm Gladwell, who had previously picked a fight with Shirky over the latter's rhapsodies about social media. Gladwell pointed out that in the 1980s, East Germans barely had access to phones, much less the Internet, and they still organized, protested, and brought down the Berlin Wall. Of the Arab Spring revolutions, Gladwell wrote, "Surely the least interesting fact about them is that some of the protesters may (or may not) have at one point or another employed some of the tools of the new media to communicate with one another."[49]

As the critics gained momentum, social media proponents fought back. Most were chastened but insisted that social media still mattered in some important way. One reporter writing for CNN hedged, "Yes, of course, technology alone doesn't make revolutions. . . . But that doesn't mean social media cannot provide wavering revolutionaries with vital aid and comfort."[50]

Amplification's Eternal Recurrence

What none of the commentators were providing, though, was a good framework for understanding technology's role, and that's where the Law of Amplification comes in. It explains how social media contributed to successful revolutions in some countries but not others, and how it can simultaneously be a supporting factor without being a primary cause.

In Tunisia and Egypt, citizen frustration and organized groups existed long before social media. Mubarak was in power for nearly thirty years, overseeing a stagnant economy under a "democracy" that had no one fooled. That frustration coalesced within existing civil society organizations and found amplified expression on Facebook. Leaders of the rebellion saw their organizing power extended by social media. Overall, technology probably accelerated the pace of revolution.

In Bahrain and Saudi Arabia, civil society was crippled, so no amount of Facebook organizing made a difference. Technology doesn't amplify human forces that aren't there.

Even Ghonim later acknowledged, "I am no hero. . . . The heroes were the ones who were in the streets, those who got beaten up, those who got arrested and put their lives in danger."[51] There is no protest without citizen frustration. There is no rebellion without a sacrifice of personal safety.

Amplification as an idea is hardly novel. My colleague Jonathan Donner, an expert on mobile phones in the developing world, pointed me to a 1970 paper on the "knowledge gap hypothesis," in which the authors reported that public-service messaging delivered through mass media was better absorbed by wealthier, more educated households.[52] Lewis Mumford, a prominent twentieth-century technology critic who was part skeptic and part contextualist, wrote a two-volume work called *The Myth of the Machine*, in which he mentions in passing that technology "supported and enlarged the capacities for human expression."[53] And Philip Agre, another computer-scientist-turned-technology-analyst, wrote prescient articles about the Internet in politics. "The Internet changes nothing on its own," he told us, "but it can amplify existing forces."[54]

But if amplification isn't new, it is completely underappreciated.

CHAPTER 3

Geek Myths Debunked

Dispelling Misguided Beliefs About Technology

In 1981 when I turned twelve, my parents gave me a Sony Walkman as a birthday present. The casing, made of brushed aluminum and a deep maroon hard plastic, glimmered as I took it out of the packaging. It was a second-generation model–light, sleek, and not much bigger than the cassette tapes it played. The headphone earbuds fit snugly in my ears, and the grooved teeth on the volume control massaged my fingertips.

That day, like hundreds of thousands of other Walkman owners, I discovered that I couldn't be without music. My biggest life concern became rationing a stash of batteries. I wanted nothing more than to spend every minute of every waking hour listening to Journey and Olivia Newton John–I still blame that Walkman for my unrehabilitated love of 1980s top-40 hits.

The Walkman poses a potential challenge to the Law of Amplification. It seems at first to be a technology that gave birth to a new human desire. Few people imagined before 1979 that they would want to live in their very own cocoons of music. Today, personal music seems to be a permanent feature of civilization. Cassette tapes have become obsolete, but headphones–and the devices they plug into–have proliferated. Didn't the Walkman change global culture? Didn't it create something fundamentally new that wasn't there before? Didn't the technology transform us in a way that we didn't previously imagine?

There's no denying that people act differently when new technologies appear. We certainly didn't walk around with tiny speakers in our ears prior to the 1980s. But that doesn't mean these new behaviors were out-of-the-blue creations of the technology, per se.

For reasons that are still not fully understood, human beings are fascinated by music. Weddings have wedding marches. Funerals have dirges. Virtuosity has been celebrated as far back as Orpheus and his lyre, and ethnomusicologists have found music in every culture, including those that ostensibly forbid it. Some traditions of Islam ban recreational music, but the Muslim call to prayer is undeniably musical. So, give people an easy way to listen to tunes–especially those of their own choosing–and it's no wonder that thirty years after the Walkman, iPods and MP3 players are still going strong. In other words, the Walkman and its descendants have allowed people to do more of something they've always wanted to do, even if that desire was never before expressed. You could call it a latent desire.

Alternative explanations hold that the Walkman caused new human behaviors. The business world uses the Walkman as a case study of a shrewd business. They say it created a new market.[1] Some sociologists argue that the Walkman changes our environment. It reorganizes space and time, what is private and what is public.[2] And as the new owner of a Walkman, I certainly felt the device beckon to me, compelling me to listen.

But statements such as "the Walkman increased sales of cassette tapes," and "the Walkman caused a portable music revolution," are shorthand for a more complex process: People have always enjoyed music, and they have personal preferences for when to listen and what to listen to. Sony leaders recognized this desire and built a low-cost, portable device to meet it. Consumers bought hundreds of thousands of units and adapted their listening habits. Other companies entered the market, expanding usage further. Throughout, it's people taking action. The device is inanimate.

It's important to keep the real explanation straight even as we use the shorthand. If we don't, we could mistakenly believe that arbitrary behaviors can be created with the right technology. We'd be tempted, by the promise of some new gadget, to, say, try to solve the problem of substandard education in America.

But technologies don't cause arbitrary behaviors. It would be easy, for example, to design high-tech clothes that make us itchy. Imagine the "Itchman" shirt made of abrasive nano-synthetic textiles and embedded with electronics that heighten static cling. If clever businesses could really create demand at will, or if technologies could cause any desired change in behavior, we could expect a smart entrepreneur to open up a worldwide market for the Itchman.[3] With today's comfort-focused materialism, though, the Itchman won't catch on anytime soon. Perhaps if we returned to a culture of penance like that of medieval times, when hairshirts were worn for repentance and mourning, we might see the rise of sackcloth fashion.

When technologies go mainstream, it's because they help scratch itches that people already have, not because they create new itches that people don't want.

FOMO and Other Four-Letter Words

Latent desires also play a role in how we use technology to connect with other people. In the age of the smartphone, many of us go out with friends and ignore each other while we tap on our gadgets. Sherry Turkle, an MIT sociologist who has studied the relationship between people and their devices for three decades, calls this being "alone together."[4] But, again, if we all seek companionship, and technology amplifies our desires, how could we be growing more apart with technology?

Some people lay the blame on imperfect technologies. Today's gizmos, they argue, provide only an impoverished form of communication.[5] You can't say much with a 140-character text message, and FaceTime is not as good as real time face-to-face. But technology doesn't necessarily block meaningful connection, either. Plenty of grandparents spend precious moments on a weekly, even daily, basis with their families over webcams. Since 2009, as many as one in five romantic relationships has started online.[6] And Facebook has done much to reconnect long-lost friends.

So it's not that technology prevents true connection. The problem is that technology also makes it easy to have thin, empty interactions. In the choice between a challenging intimacy and casual fun, some of us choose the latter. One reason why some people can't stop fiddling with their phones is something called FOMO–"the fear of missing out."[7] The fear of missing out on a better party, a better evening, a better life.

But again, the technology doesn't cause this behavior; it just amplifies the underlying personality, turning us into caricatures of ourselves. I have friends whose handsets I've never seen. When we sit down for a meal, their phones stay in their purses or pockets. If there's a ring, they ignore it. At the other extreme, I have acquaintances with whom conversation devolves to a few words between interruptions. Even when they're not texting, their fidgety glances land on their phones like mosquitos seeking a soft patch of skin. Over time, I've come to see FOMO as just one of many causes of smartphone obsession. There's also ATUS, addiction to useless stimulation; PORM, pleasure of receiving messages; SWAP, seeing work as priority; UTSI, the urge to seem important; and any number of other latent emotional tics that are exacerbated by the technology. The fact that owners of the same kind of device display a diverse range of behaviors is another sign that the technology is amplifying what's already there, not causing the same response in everyone.

Commentators have thought hard about what made the Walkman and the iPhone such successes, but no one asks why there's no mass market for the Itchman. This lopsided focus on technologies that "succeed" blocks us from seeing the full picture. It's like the pundits who forgot Bahrain and Saudi Arabia. Claims of the Internet's democratizing power fail to take into account the many things that the Internet hasn't democratized, such as wealth, power, and genius.

When evaluating theories of technology, we should look at a wide range of contexts. Our conclusions should come not just from isolated instances or personal experience, but from all kinds of uses in all kinds of circumstances. In this chapter, I examine a smattering of examples: from electronic medical records to corporate knowledge management, from politics in America to Chinese media censorship. It won't be comprehensive. But out of these scattershot case studies, a pattern will become clear.

Along the way, I'll also demolish a few persistent myths. It's often said that technology is a cost-saver; or that "big data" makes business problems transparent; or that social media brings people together; or that digital systems level playing fields. These kinds of statements are repeated so often that few people question them. Yet none of them is a die-cast truth.

If Hippocrates Were an Economist

One of information technology's great benefits, supposedly, is its ability to lower costs. Walmart, for example, is famous for its digital stock-keeping. Its databases know exactly what's on the shelves, and they automatically inform suppliers which stores are low on stock. The system keeps inventories razor-thin and costs low. And it all seems to be about technology–databases, barcode readers, RFID-encoded pallets, and so on.

You might think, then, that some of our greatest cost-control challenges could be solved with IT. A conspicuous target in America is our health-care system. In fact, electronic medical records have firm bipartisan support even in an era of political deadlock. President Barack Obama has called for electronic medical records since before his days in the White House, citing efficiency and cost savings.[8] And the GOP Doctors Caucus, formed by Representatives Phil Gingrey and Tim Murphy, states, "Health information technology has the potential to save more than $81 billion annually in health care costs. From drastically reducing medical errors to streamlining administration, health IT is the key to transforming our healthcare system."[9]

Unfortunately, cost containment also follows the Law of Amplification. In the American health-care system, very few people are really focused on reducing costs. As a result, every new technology is a white elephant–a "gift" you have to keep paying for. Many of us, sadly, are familiar with this state of affairs. A few years ago, I went to see a specialist in neuro-ophthalmology because I had lost partial vision in my right eye. After asking me some questions and peering into my pupil, the doctor said, "Well, there's no clear problem, so it could be nerve damage. If it is, there's not much we can do. But," he said, smiling conspiratorially, "since you have good insurance, let's do an MRI." I agreed because I had no reason not to. I was lucky to have insurance with no co-pay. When I saw the invoice for the visit, the line item just for the MRI showed $1,800. I was shocked, though grateful that my insurance covered the expense. The doctor's office never called me for a follow-up, the MRI scan was never consulted, and my right-eye problems persist.

Unlike at Walmart, where digital tools amplify the company's zealous pursuit of lower costs, in US health care, technology intensifies all the ways in which spending is encouraged. Our hypochondria as patients, our foibles as doctors, our greed as suppliers, and our myopia as policymakers–all are social forces that the technology regrettably amplifies. Even the employers and governments that foot the bill cast their payments as benefits to employees and citizens. They don't penny-pinch, for fear of appearing cavalier about people's lives. On top of everything else, our metrics are off. As Princeton University economist Uwe Reinhart noted, "Every dollar of health care spending is someone's health care income."[10] That income flows right into our national gross domestic product (GDP), and we want the GDP to rise, don't we?

Of course, technology also amplifies good health-care trends, and that's terrific for those who can afford it. But if lowering costs is the goal, more technology isn't a surefire solution. In the four decades since 1970–a period during which digital technologies poured into hospitals and clinics–American health-care costs rose in real terms by a factor of five. The increase has been far greater than in other developed countries.[11] Information technology was probably not the main cause, but it certainly didn't turn the tide. (Nor did we get what we paid for: American life expectancy during that period only increased by eight years. That's fewer than the nine years gained in the United Kingdom and the eleven gained in Japan, even though they spent a lot less.[12])

So lower costs aren't a function of the technology itself. If anything, digital technologies require additional upkeep. For example, in 2010 when I left Microsoft, the firm employed over 4,000 full-time people to keep its own IT systems running. That's nearly 5 percent of the company's workforce. (Similar proportions hold for any large technology company.[13]) If technology companies–which work hard to automate everything–have to spend 5 percent of their human resources managing IT, imagine how much more difficult it is for other organizations.

Especially in the context of US health care, digital tools just amplify what is already an outrageous system of accounting. Recent exposés show that patients are billed excessive prices routinely: $24 for a niacin tablet that comes to 5 cents at drug stores; $333 for a chest x-ray costing less than $30; $49,237 for a neurostimulator that wholesales at $19,000 and might cost only $4,500 to manufacture.[14] In this climate, hospital administrators will be happy to install electronic medical records and pass on the costs to patients and taxpayers at a markup.

Managing "Knowledge Management"

So, contrary to popular belief, digitization by itself doesn't necessarily reduce costs. What about improving organizational behavior? Won't computers solve our knowledge-management problems? And won't "big data" make traditional decision-making obsolete?

As with cost cutting, information technologies *can* improve knowledge exchange and transparency, but they don't do so automatically. Curiously, a group of people who I thought would resist that message turned out to be sympathetic. They understood exactly what I was saying.

One of them was Jorge Perez-Luna. He's held titles like chief information officer and VP of IT at telecommunications companies including AT&T, Motorola, and Nextel. At one company, Perez-Luna was asked by his CEO to implement a computerized order-tracking system for their sales office in Brazil. The office was consistently underperforming, and the boss wanted a fix. He thought that by installing a database to track sales, he could find the problem.

Perez-Luna sent a small team on an exploratory mission. They found that "one employee had a drawer full of signed contracts, all of which were uncollectibles"–outstanding payments due from customers. "And he wasn't an exception." It turned out that salespeople were rewarded for signing deals, but they didn't follow up with customers. Managers set quotas for new contracts, but there were no processes in place to handle uncollectibles. The sales staff didn't know how much income they were bringing in, and they didn't have any reason to care. As Perez-Luna put it, "payments to the company just weren't being prioritized."

Perez-Luna reported back. Without plugging that managerial hole first, he told his boss, technology wouldn't do much good. He recommended more oversight and a shift in priorities. Not only had he saved the company money by avoiding an expensive digital solution, but he had identified the true problem. "I'm an IT guy," he said, "but some of my best friends have training in anthropology. They are good at seeing the human issues behind technology."

New laptops don't necessarily make employees more productive. State-of-the-art data centers don't cause better strategic thinking. And knowledge-management systems don't cause rival departments to share information with one another. Yet CIOs everywhere are asked to perform exactly that sort of wizardry. The more experienced ones are careful not to promise too much. Technology can improve systems that are already working–a kind of amplification–but it doesn't fix systems that are broken. There is no knowledge management without management.

In large organizations such as universities, governments, and corporations, one hand frequently doesn't know what the other hand is doing. To break down silos, it's tempting to set up Web portals and internal social media sites, but the real issues are almost always those of management, internal politics, and even limited human attention. Unless those social problems are dealt with, technology doesn't have a base to amplify. Especially in a world where everything is already digitized, knowledge-management systems and online clearinghouses are rarely the bottleneck. To clear organizational obstacles, the counterintuitive solution in an age of bountiful technology is to focus on building effective human relationships.

Reach Out and Touch Your Tribe

Speaking of relationships, technology is often believed to enhance them. Nokia's tagline is "Connecting People," and AT&T once used the slogan "Reach out and touch someone." There's no doubt that communication technologies help people connect, but there are at least two ways in which this could happen. Option A says that better tools help us communicate with people we are already inclined to communicate with. Option B says that better tools cause communication to occur where none previously existed or was desired.

Amplification votes for option A: We use new tools to communicate more with people we want to connect with anyway. A host of evidence supports this conclusion. For example, a Pew Research Center study shows that, on average, about 92 percent of our Facebook friends are real-world acquaintances, not random people we've connected with because of the Internet.[15] Other studies show that people collaborate more with those they are physically close to already.[16] Despite email and Twitter, a single flight of stairs between offices can inhibit working together. All of this is to say that we use electronic communication to strengthen–amplify–preferred relationships.

Option B leads to the misguided belief that more connectivity brings *everyone* closer together. As one utopian put it, "People will communicate more freely and . . . the effect will be to increase understanding, foster tolerance, and ultimately promote worldwide peace."[17] This may sound horribly naïve, but the author, Frances Cairncross, is hardly an intellectual lightweight. She has been a journalist for *The Guardian* and *The Economist* and has held top posts at Britain's Economic and Social Research Council as well as the British Science Association.

It's easy to see that more communication tools don't lead to better relationships or mutual understanding where neither previously existed. Consider that the United States has never before had as many communication options as it has now. In the 1970s, having a television meant access to ABC, CBS, NBC, and maybe a staticky PBS. Today it means cable service, Internet streaming, and access to hundreds of channels. In the 1970s, most households had landlines, but only the Yellow Pages for one town or city. Today you can look up just about anyone online and call them on the move. In the 1970s, only the geek elite used email. Today everyone texts, tweets, and posts to Instagram. Yet none of this extra connectivity seems to be bridging the chasm between the political left and right. If anything, the gulf is widening.

What is actually happening was predicted by MIT professors Marshall Van Alstyne and Erik Brynjolfsson as early as 1996–two years before Google and eight years before Facebook. "Internet users," they wrote, "can seek out interactions with like-minded individuals who have similar values" while minimizing interactions with those whose values differ.[18] Van Alstyne and Brynjolfsson called this phenomenon "cyberbalkanization"; psychologists call it "selective exposure."[19] Online, you can find self-reinforcing groups of white supremacists on the one hand, and free-loving hippies on the other. And the effect goes well beyond the Internet. Thus liberals watch Jon Stewart, while conservatives watch Glenn Beck. Gone are the days when Americans all tuned into Walter Cronkite and heard the same news with the same commentary. The danger of cyberbalkanization is that people become radicalized, intolerant, and "less likely to trust important decisions to people whose values differ from their own."[20]

It's true, of course, that communication tools can bring people closer. Olympic broadcasts help unify countries with pride. In the week after I got on Facebook, I happily connected with friends from the third grade. But these are examples of people using technology to do more of what they already want to do, not making friends with old enemies.

How Not to Bridge the Digital Divide

The target of many social causes is some kind of inequality–of wealth, education, political voice, social status. Another is the "digital divide," a phrase coined in the 1990s to describe unequal technology access between rich and poor Americans. The term was quickly extended to global disparities, and soon bridging the digital divide became a rallying cry. One response was to develop low-cost technologies–to make things only rich people could own affordable for everyone. This was the idea behind One Laptop Per Child. Its early media buzz was based on a projected $100 price tag.[21] The Indian government rejected OLPC and proposed instead its own low-cost tablet, the Aakash, for $35.[22] And as early as 1999, there was Free-PC in the United States. The company offered PCs for $0; they were paid for by on-screen advertising.

Free-PC was discontinued, and the other products never hit their target prices. But bad business models aren't the real problem with these efforts. The problem lies in the concept itself. Some people speak of low-cost access to goods as a kind of "democratization," but in a real democracy, it's one person, one vote. In a free market, it's one dollar, one vote, which is a totally different beast. Richer people can always afford more technology. It's not as if new technologies stop appearing while existing ones are made cheaper. By the time there are low-cost PCs, there are high-cost smartphones. By the time there are low-cost smartphones, there are high-cost phablets.[23] And by the time there are low-cost phablets, there will be high-cost digital glasses. There is no technological keeping up with the Joneses.

But suppose that an even distribution of technology were actually possible. What then? To answer this question, consider the following situation. Imagine the poorest person you can think of who is involuntarily poor. (The involuntary part is important–I'm not asking you to imagine a contented monk.) It might be a homeless person in your city, or a poor migrant worker in a remote area. Now imagine that you and that person were asked to raise as much money as you could for a charity of your choice, using nothing other than unlimited access to email for one week. Who would be able to raise more money? For most readers, it will be you. Because you have richer friends. You probably have more education and can write more persuasive emails. You likely have better organizational skills and could rally more people to the cause. And depending on the poor person you imagined, you might also be far ahead in basic skills such as literacy.

In this thought experiment, the technology is identical, but the outcome is different because of what you each started with. The differences are all about people–who you are, whom you know, and what you're capable of. These are the same factors, incidentally, that allow you to be richer in the first place. Imagine repeating the same experiment, but not with someone who's poor. Do the experiment with Bill Clinton or Bill Gates. Who would be able to raise more money, you or one of them? One of the Bills would, for the same reasons.

You could repeat these experiments with different information technologies (e.g., mobile phone calls, Twitter) and with different tasks (e.g., finding a job for your friend, seeking investment advice), and, for the most part, the results would be the same. In each case the technology is fixed, but the outcomes differ in proportion to the underlying advantages. Low-cost technology is just not an effective way to fight inequality, because the digital divide is much more a symptom than a cause of other divides.[24] Under the Law of Amplification, technology–even when it's equally distributed–isn't a bridge, but a jack. It widens existing disparities.[25]

The Chinese Elephant

Harvard political scientist Gary King, who studies, among other things, the Chinese Internet, says it is the site of the "most extensive effort to selectively censor human expression ever implemented." King has uncovered exactly what the Chinese government censors on its country's social media platforms, and what he has found has unexpected lessons far beyond the digital realm.[26]

According to King, "the Chinese Internet police force employs an estimated 50,000 censors who collaborate with about 300,000 Communist Party members. In addition, private firms are required by law to review the content on their own sites," and for this they hire staff. King has reported that the overall censorship effort is so large that "it's like an elephant walking through a room." To track and measure its footprints, he conducted two subversive studies with colleagues Jennifer Pan and Margaret Roberts that offer new insights into the Chinese Leviathan.

In the first study, the team built a network of computers that watched 1,382 Chinese websites, monitoring new posts to see if and when they were censored. Eleven million posts covering eighty-five topics were chosen for investigation. The subjects ranged in political sensitivity from popular video games to the dissident artist Ai Weiwei. The researchers included online chatter resulting from real-world events.[27] In the second study, King and his team went undercover. They created fake accounts on over one hundred sites. They submitted posts to see which ones were censored. They even set up their own social media company in China.[28]

Two of their findings stand out. First, China's online censorship mechanisms are panoptic and efficient. Objectionable items are removed with a near-perfect elimination rate, typically within twenty-four hours of their posting. The researchers wrote, "This is a remarkable organizational accomplishment, requiring large scale military-like precision."

Second, King and his team found what Chinese censors don't like. They're quick to act on anything that refers to, instigates, or otherwise links to grassroots collective action. Posts about protests, demonstrations, and even apolitical mass activities vanish quickly.[29] But the regime is comparatively comfortable with criticism of the government. For example, this passage was not censored:

> The Chinese Communist Party made a promise of democratic, constitutional government at the beginning of the war of resistance against Japan. But after 60 years that promise has yet to be honored. China today lacks integrity, and accountability should be traced to Mao. . . . [I]ntra-party democracy espoused today is just an excuse to perpetuate one-party rule.

Meanwhile the following post, which refers to a man who responded to the demolition of his home by carrying out a suicide bombing, was nixed:

> Even if we can verify what Qian Mingqi said on Weibo that the building demolition caused a great deal of personal damage, we should still

> condemn his extreme act of retribution. . . . The government has continually put forth measures and laws to protect the interests of citizens in building demolition.

This comment was supportive of the government, but it was censored because it referred to a known source of public agitation. The distinction contradicts conventional ideas about totalitarian states. In George Orwell's *1984*, Big Brother dealt quickly with any expressed disloyalty. But King's findings reinforce more subtle theories of autocratic power, like that of his colleague Martin Dimitrov, who has argued that "regimes collapse when its [*sic*] people stop bringing grievances to the state."[30] The real danger to a state comes when its citizens no longer complain in the open.

In fact, as King noted, a certain amount of public criticism may serve the Communist Party's interests. It mollifies citizens who want to blow off steam, and it alerts the central government to issues requiring attention. It's when the criticism spills over into calls for action that the censorship machine–and sometimes also the police–kicks in. The government is continually calibrating its tactics. In October 2013, a man in Shaanxi Province was detained for having a critical comment re-tweeted 500 times on Sina Weibo, China's version of Twitter.[31] You can almost hear bureaucrats debating where the line should be: How many shares pose a collective-action threat–250, 500, 1,000?

King's study of Chinese social media censorship, then, reveals a lot more than just a strategy for online speech suppression. It provides clues to the Communist Party's deepest fears and its sophisticated program of control. As we've seen with its heavy-handed response to uprisings in Xinjiang and Tibet, China is serious about suppressing physical protest. That intention carries over online, where censors are sensitive even to seemingly innocuous posts if they contain a seed for mass action. In a phone conversation, King told me that "political actors in any country use whatever means of communication they have to advance their goals. If technology allows them to do it faster, they'll use technology."

"In some ways, it's the same in America," King continued. Indeed, large technology companies in the United States are legally required to monitor and censor illegal content such as child pornography. And we know from recent revelations about the National Security Agency that our government is willing to strong-arm firms for the purposes of digital surveillance. "Functionally, that's the same as what happens in China, though I won't say it's morally the same," King said. In both countries, technology acts like a lens, magnifying and amplifying how governments act on their gravest concerns. By examining large-scale technology, you can ferret out hidden motivations.

Predicting Is Believing

The Law of Amplification enables us to make certain types of predictions. Under some conditions, it's possible to gauge the future of a technology that doesn't even exist yet. For example, imagine that scientists come up with the following inventions. In each pair, which one do you think would be more popular?

a) A robot that cleans up after you, washes your dishes, and does all of your laundry.
b) A robot that follows you around and verbally points out each of your personal flaws.

a) A holographic device that projects the realistic illusion that your house is bigger than it is, outfitted with expensive furniture, and decorated by a professional interior decorator.
b) A holographic device that projects the realistic illusion that your house is smaller than it is, outfitted with used furniture, and decorated by a college student.

a) A novel device you wear on your belt buckle that guarantees a slim, fit figure, regardless of what you eat or how much you exercise.

b) A novel device you wear on your belt buckle that guarantees an overweight figure, regardless of what you eat or how much you exercise.

None of these devices exists today, but you will have no trouble picking which of each pair would sell better. That's because you already have a good sense for what most people want. Your ability to predict a technology's success is based on an intuitive grasp of the human condition. Consistent with amplification, human preferences, more than technological design, decide which products succeed. Or, to put it another way, good design is the art of catering to our psyches.

You might quibble about which way these options would go. You might say that the outcomes depend on culture or the moment in history in which they occur. And you'd be right. What many Americans now consider an undesirable weight has been in other times and places a sign of wealth and status–for example, in the time of Peter Paul Rubens, who painted what we now call Rubenesque women.[32] Back then, device (b) would have done better than device (a). But that again proves that the technology doesn't decide its outcome.

Similarly, we can predict that in future revolutions, all sides will use or abuse the communication technologies at their disposal. In the nineteenth century, rebels distributed pamphlets, autocrats closed printing presses, and the world heard about it months later by word of mouth. Here in the twenty-first century, rebels organize on Facebook; autocrats shut down the Internet; and the world watches events unfold on YouTube. Perhaps in the twenty-third century, rebels will rally on brain-to-brain transmitters; autocrats will scramble neuro-signals; and the world will watch it all through their synaptically projected awareness modules (known in the future as "SPAM"). The digital world is undoubtedly different from the analog and the postdigital, yet for so much of the social order . . . *plus la technologie change, plus c'est la même chose.*

Most importantly, amplification provides a guide as to whether social-change dollars should be spent on undeveloped technologies or on something else. We've seen how struggling schools aren't turned around by digital technologies, but tech utopians will insist that the right technology just hasn't been invented yet. So let's entertain their reverie for a moment and imagine a world with a powerful teaching machine like that from *The Matrix*.

I plug myself in, and, within seconds, "I know kung fu," just like Keanu Reeves's character in the movie. It's an amazing technology that could teach just about anything, but will it eliminate inequities in education? In any world politically like ours, wealthy, influential parents will secure the best hardware for their own children, while the children of poor, marginalized households will have access to older models in need of repair. Rich kids will effortlessly learn quantum physics. Poor kids might come out quacking like a duck. Yet again, the technology will amplify the intentions (both explicit and implicit) of the larger society. And the same will be true of gamified e-textbooks, humanoid teaching robots, or any other novel technology. So, one prediction is this: If you're interested in contributing to a fair, universal educational system, novel technology isn't what will do the trick.

The broader lesson applies well beyond education and summarizes what we've seen so far. A government without genuine motivation to eradicate corruption will not become more accountable through new technologies of transparency. A health-care system with a shortage of well-trained doctors and nurses won't find its medical needs met with electronic medical records. A country unwilling to address the social underpinnings of inequality won't see an end to inequities regardless of how much new low-cost technology it produces. In general, technology results in positive outcomes only where positive, capable human forces are already in place.

In Chapter 6, I'll show how all of this offers guidance for the best use of technology, but for now, let me mention that the Law of Amplification's predictive power is one of its strengths as a theory. Want to know where free speech is most likely to thrive online? It will be where it thrives offline. Want to know when new technology will actually cut costs? It will be when management is focused on cost control. Want to know how to ensure that your children will learn productively on an iPad? It will be if they have good learning habits independent of the tools at their disposal and adult guardians monitoring proper use.

Whither the Unintended Consequences?

Some people might protest that technology outcomes are fundamentally unpredictable because of unintended consequences. They'd be right up to a point, but only up to a point. Nothing I've said so far says that that human history is more predictable just because it involves technology. Who knew in 2010 that the Middle East would be transformed by popular uprisings, with or without Facebook? People are complicated and hard to predict; adding technology doesn't change that.

But where social situations are well understood, some technology outcomes can be predicted, and even partial or imperfect knowledge is valuable. Most of us consult weather reports even though we know they're sometimes wrong. Any information that is better than a random guess is still useful. Similarly, superintendents can act on predictions that educational technologies will help good schools but not struggling ones, even if borderline cases are harder to assess.

In addition, whether a consequence is unintended often is in the eye of the beholder. Officials at the US Department of Defense or the National Science Foundation who sponsored precursors of the Internet probably didn't mean to pave the way for either mass electronic commerce or the global proliferation of cat videos. So it could be said that those outcomes were unintended. But websites don't just build themselves. Everything online is someone's intent acted out, even if those contributions weren't foretold by Internet founders. More often than not, the unintended consequences of technology spring from someone else's unpredictability. One man's unintended consequence is another man's mission.

But what about cases in which a technological result was absolutely impossible to foresee? Pure examples are hard to find, because technological skeptics have vivid imaginations about what can go wrong.

But for the sake of argument, consider teenage texting. It seems unlikely that either the engineers behind the SMS text-messaging standard or the parents who welcomed mobile phones into their households imagined that their children would one day send thousands of text messages a month. Yet, on average, American teens send and receive 60 texts a day, or 4 per waking hour. One 13-year-old girl in California exchanged 14,528 texts in a month, which comes to about 1 every 3 minutes, 24 hours a day.[33] So obsessive texting could be considered an unintended consequence of mobile-phone proliferation.[34] But now that we know about it, it's no longer unintended. It's up to adults–as parents and consumers, as voters and citizens, as nuclear families or as collective communities–to decide whether this consequence is desirable, and, if not, to curtail it. To do nothing is to be complicit in–to passively intend–the undesired outcome. In the long run, there are no unintended consequences.[35]

Deus In Machina

The Law of Amplification explains how technology can be both good and bad, and how its effect is ultimately up to individuals and societies. The law's corollaries dispel myths about technology's inherent powers, whether to lower costs, improve organizations, or decrease inequality.

Amplification also pegs the responsibility for technology's impact squarely on us. Techno-utopians see a world where technology saves us from ourselves. Cyber-skeptics imagine our creations running rampant. And contextualists often sound like apologists for luck. All of these views, however, smack of humanity's naïve youth, when we thought our lot was up to the Fates, to nature, or to God. Both excessive faith in and frantic fear of technology are regressions to childhood, denials of human responsibility. In our post-existential adulthood, shouldn't we own our destinies?

To adapt Jean-Paul Sartre, technology is nothing else but what we make of it.[36] And as Sartre noted, that responsibility is both a blessing and a curse–on the one hand, we *can* decide what to do with technology; on the other hand, we *must* decide what to do.

CHAPTER 4

Shrink-Wrapped Quick Fixes

Technology as an Exemplar of the Packaged Intervention

Over the years, I've given a lot of talks about digital technology's limited impact on social causes, and I've heard a range of reactions. Tech champions respond with hostility or begrudging resignation. Those tired of the hype offer commiseration. And another group of people say they agree, but not with respect to electronics. They say they've had similar experiences in nontechnology projects.

Amplification, it turns out, applies more broadly. Digital technologies are just an extreme example of what could be called the *packaged intervention*–any technology, idea, policy, or other easily replicable partial solution meant to address a social problem.[1] Like technologies, they're expected to cause large-scale social change in and of themselves. But they don't, and understanding why provides further insight into the Law of Amplification.

The Technology of Microcredit

A good example of a packaged intervention is microcredit. Small-scale lending to poor borrowers, also called microlending or microfinance, is one of the few ideas in poverty alleviation to have attained global reach. There is no end to the tales of $100 loans transforming the lives of low-income households. Muhammad Yunus, the grand patriarch of modern microfinance, likes to tell one such anecdote. He once made a personal loan of $27 distributed among forty-two villagers in Bangladesh. Some of them needed the tiniest bit of cash to buy materials to make bamboo stools. They could pay back the loan and still earn a profit. "I had never heard of anyone suffering for the lack of *twenty-two cents*," Yunus wrote.[2]

That was 1976. In the decades since, microcredit has grown into a titan of social programs. It was boosted by the tireless evangelism of advocates like Yunus, who has said that "the purpose of microcredit is to eliminate poverty in the shortest possible time frame."[3] According to proponents, loans allow households to apply existing skills to build up small businesses–microenterprises–and climb out of poverty on their own. Microcredit is supposed to increase incomes, empower women, and enhance health and education outcomes for children. It's been said that lives are "transformed by microfinance," and that microcredit "revolutionized global anti-poverty efforts."[4] Today there are as many as 180 million microcredit borrowers worldwide.[5] The United Nations declared 2005 the International Year of Microcredit. In 2006, Yunus won the Nobel Peace Prize.

But, as even advocates concede, credit is not a panacea. Some microfinance organizations prioritize growth and profit. From there, it's a slippery slope to abuse. One of the most controversial lenders is Compartamos Banco in Mexico, which started as a nonprofit but reconfigured itself as a for-profit bank so that it could raise commercial capital. In 2007, after demonstrating an incredible average return of 53 percent over seven years, the bank raised $467 million in an initial public offering (IPO).[6] Of course, to provide such a high rate of return, Compartamos charges even higher interest rates to its borrowers. Many of its loans have an annual percentage rate (APR) of over 100 percent–well above what is allowed by state usury laws in America, and much higher than US credit-card rates, which rarely go above 30 percent.

Eight months after the IPO, *Bloomberg Businessweek* tracked down a woman named Eva Yanet Hernández Caballero, who appears in Compartamos's promotional materials as a success story. To purchase material for a sock-knitting business in 2001, Hernández had taken out loans of up to $1,800 with an APR of 105 percent. When her customers fell behind on their payments, though, she fell behind on hers. Hernández was thrown out of her borrowing circle and disqualified for further loans. She said her family was "barely afloat," even as her photo contributed to the bank's glow. Compartamos claimed that interest rates must rise in order to cover costs and attract investors. But with some of its early investors walking away with hundreds of millions of dollars while people such as Hernández struggle, it's hard to interpret their work as anything other than profit-seeking.[7] In a *New York Times* op-ed, Yunus, whose Grameen Bank charges only 20 percent interest and refinances loans on favorable terms when borrowers can't make payments, blasted Compartamos: "Commercialization has been a terrible wrong turn for microfinance. . . . Poverty should be eradicated, not seen as a money-making opportunity."[8]

Profits might be forgivable if borrowers always benefited. Recent research, though, tempers microlending hype. For one thing, microcredit is usually just one of many credit options. One insightful set of studies based on financial diaries of households in Bangladesh, India, and South Africa shows how poor families adroitly juggle informal loans, savings, and insurance.[9] And, contrary to common claims, borrowers don't all build businesses. More often, loans are used to pay off other debts or to absorb financial shocks. For instance, people borrow for medical crises in a kind of post-hoc insurance called "consumption smoothing." In other words, sometimes microcredit is less about moving out of poverty and more about basic survival.

Leading development economists have also cast doubt on microcredit as a path out of poverty.[10] One of them is my friend Dean Karlan, a Yale professor whose work takes him on regular odysseys around the globe. (His youngest daughter had a platinum frequent flyer card when she was eight years old.) Karlan and a colleague, Jonathan Zinman, found mixed results with small loans in South Africa and the

Philippines. On the positive side, among moderately poor adults in South Africa living on $6 per day, borrowers were more likely to end up with jobs, more likely to rise above the poverty line, and less likely to experience hunger in the family.[11] So far, so good. But in the Philippines, loans made to active microentrepreneurs "did not generate bigger businesses, higher income, and higher subjective well-being, but rather led to stronger risk management, fewer businesses, and lower subjective well-being."[12] And in a study of Compartamos Banco with colleague Manuela Angelucci, they concluded, "On average, businesses do not become more profitable and income does not increase, but people are better able to manage debt without selling assets, women gain more power in the household, and happiness increases."[13] In other words, microcredit offers some benefits, but it isn't a cure-all for poverty.

These results caused a commotion in the world of microfinance. They disputed the claims of revolutionary, life-transforming outcomes. The findings on the whole favored microcredit, but not overwhelmingly, and they were interpreted as lukewarm. Another group of researchers who found similar results put a provocative question mark in their paper's title: "The Miracle of Microfinance?"[14] The industry took a defensive crouch, and supporters of small loans hit back with real-life rags-to-middle-class stories and accusations that the investigators only looked at short-term impacts.

For now, something of a truce prevails. Both sides agree that microcredit can be helpful, that the more extreme claims are overblown, and that more research is needed.

An Economic Trojan Horse?

Credit can be good or bad wherever you are in the world. In the United States in 2006, for example, there were two credit cards for every person, and the average card carried $4,800 in debt.[15] Median household debt was around $70,000, and the personal savings rate hit–1.1 percent, the minus sign denoting that Americans were spending more than they were earning.[16] Much of this debt was in the form of subprime home mortgages. Banks marketed them aggressively to homebuyers as affordable loans, then repackaged them as opaque, high-interest securities for investors. Mass defaults on the mortgages have been widely cited as the trigger of the 2007–2008 financial crisis from which the global economy is still reeling.[17] Credit, clearly, is a powerful but dangerous instrument. Making it more available doesn't always lead to good outcomes.

Good credit depends on at least two things: the care with which the loans are made, and the economic capacity of the borrowers. This is true of any loan. Grameen Bank is known to take immense care with its loans. Many of its borrowers are undoubtedly benefiting.

But which ones? A closer look at the economists' data shows that positive effects tend to favor certain subgroups. Microcredit is more beneficial for those with greater wealth and education; for those with existing businesses; and for those with entrepreneurial skills and temperament; and, in some communities, for the men more than for the women, probably because of other sociocultural advantages.[18] In other words, like digital technologies, microcredit also amplifies human forces.

But microcredit isn't a concrete thing like hardware or software. What is doing the amplifying? In essence, it's a prescribed process for making cash loans. In its classic form, microcredit is based on lending to groups of people who offer a social guarantee instead of financial collateral. Formal banks make loans to the whole group, often mediated by microfinance institutions. The group then distributes smaller loans to individual members. Borrowers support one another in making loan payments and apply peer pressure to ensure that individual members pay.

By bundling a method of small-scale loan-making into a standard package, microcredit is like a technology.[19] As a packaged intervention, it has spread far and wide. Its reach, though, outpaces its positive impact. When microloans are promoted as ends in themselves, essential elements of pioneering programs such as Yunus's are often forgotten. Those elements, I have come to believe, are the same ones we neglect when we invoke technology to solve a social problem.

Democracy in a Box

Packaged interventions come in all shapes and sizes: iPads to supply children's education; condoms to prevent the spread of HIV/AIDS; and you may have seen ads from organizations such as Oxfam and Heifer International asking you to donate a goat–great as food and fertilizer source for a poor farming family.

But, as with microcredit, packaged interventions aren't limited to physical goods. They can be abstract ideas or institutional structures: school vouchers, charter schools, home mortgages, elections.

Elections are hailed as the means to achieve democracy, and US foreign policy seems fixated on having other countries hold them. Few events provoke the media frenzy of a nation's first election.

Recent events in the Middle East and Afghanistan, though, remind us how little voting accomplishes in and of itself. Tunisia, Egypt, and Libya overthrew their dictators and held elections for new leaders, but it's far from clear that democracies will follow. Egypt, for example, elected Mohamed Morsi in its first elections after the 2011 revolution, but Morsi was subsequently ousted in a military coup. Libya broke apart into fiefdoms controlled by regional militias. Meanwhile, even when the world's strongest military power ousts despots and supports elections–as America has done in Iraq and Afghanistan–the results are often corruption and violence.[20]

Most political scientists and foreign policy wonks shudder at glib faith in elections. They point out that democracy requires much more than voting: the rule of law, a robust free press, widespread education, strong governing institutions, public demand for government responsiveness and transparency, civilian control of the military, moderate levels of wealth without too much inequality, and so on.[21] These things are not easy to package, as they depend on prevailing social norms, the personalities of leaders, and other human factors that are hard to replicate, even at gunpoint.

A notable comment along these lines comes from Chinua Achebe, the celebrated Nigerian author of *Things Fall Apart*. Achebe was known for his dislike of Western condescension, but he still acknowledged the long road ahead for true Nigerian democracy.[22] In 2011, he wrote, "Africa's postcolonial disposition is the result of a people who have lost the habit of ruling themselves, forgotten their traditional way of thinking, embracing and engaging the world without sufficient preparation." He also wrote, "Restoring democratic systems alone will not, overnight, make the country a success. . . . We also must realize that we need patience and cannot expect instant miracles."[23]

Fellow Nigerian and Pulitzer Prize winner Dele Olojede went even further: "If you have a large number of citizens who do not yet fully grasp even the concept of the state, who are largely very poor and uneducated, who struggle daily to survive and [are] susceptible to a $5 bribe, democracy sometimes seems like a sham." Olojede even wondered about giving every adult the vote, noting that no "reasonably successful democracy" began with universal suffrage.[24]

Elections don't guarantee political freedom, accountable governance, national stability, and citizen well-being. Like other packaged interventions, ballot boxes are relatively easy to replicate, but a healthy democracy requires much, much more.

Equality by Decree

A group experiences social, political, and economic discrimination within the country of their birth. Its members suffer from limited job opportunities, physical segregation, and biased restrictions on marriage and family life. Then, with great fanfare, laws are passed to end the prejudice. Yet, while the laws have an undeniable effect, bigotry and social inequities continue for generations.

That is how just about every fight for social and political equality goes.

It's the story, broadly, of America's quest for racial and gender equality. In the early 1960s, the United States passed the Equal Pay Act and the Civil Rights Act. The first outlawed wage disparities by sex, and the latter proscribed a range of discriminatory practices based on race, creed, gender, ethnicity, and national origin. It's been over half a century since then, and yet, while outright misogyny has decreased, the gender wage gap hasn't closed.[25] And racial inequality hasn't gone away despite the election of a black president.

Enduring bias also plagues India. Many of the rural villages I visited were divided into hamlets separated by caste. In some of these villages, strict norms govern physical interaction, some of which recall the rules of kindergarten cooties. For example, people of the *dalit* outcaste are often forbidden to touch other residents, to drink water from the same wells, or even to allow their shadows to fall on upper-caste neighbors. Infringements are met with beatings for the offender and cleansing rituals for the Brahmin.

Modern Indian urbanites often claim not to engage in such ostracism, but that's true only in a relative sense. From women living in slums, I learned that many upper-middle-class families refuse to hire cooks who aren't of a particular caste. In elections, there's a saying that "people don't cast their votes; they vote their caste." And in the personal stories I heard at my office, the most common dramas recapitulated *Romeo and Juliet*, with the Montagues and Capulets replaced by rival castes.

Caste discrimination is eroding in India, and some of the shift can be attributed to law. But considering that India's constitution has prohibited untouchability and discrimination since at least 1950, caste-ist practices are woefully persistent. In much of the country, especially in rural areas, little has changed.

In other words, even laws and policies are a kind of packaged intervention–easy to replicate on a national scale, but still unable to achieve their stated goals on their own.

The Ultimate Package: Vaccines

In any pantheon of packaged interventions, vaccines would reign supreme. They save lives. They work quickly. They don't need any follow-through. And they're so effective that other packaged interventions have vaccine envy. Of One Laptop Per Child, Negroponte has dubiously said, "Think of it as inoculating children against ignorance. And think of the laptop as a vaccine."[26]

Vaccines are a bit of medical magic, as long as they're successfully delivered to people who are willing to receive them. Polio has been eradicated in rich countries, but it's still endemic in Afghanistan, Pakistan, and Nigeria, and it smites victims in a dozen other African countries. Measles continues to claim over 100,000 lives a year in the developing world, with large outbreaks in Asia and Africa. Why is it that vaccines fail to protect so many, even though they are an ideal packaged intervention?

Could the problem be a lack of vaccine technology? Obviously not—vaccines for polio and measles have existed for decades.

Could it be that the technology is too expensive? It's true that countries struggling with these diseases can't spend as much on health care as developed nations do. But the oral polio vaccine, for example, costs no more than 20 cents a dose. It's affordable even for poor countries. To supply the recommended three doses for all 350 million people living in polio-endemic countries would cost $210 million. That's not a small sum, but it's well below the nearly $1 billion budgeted in 2011 alone by the Global Polio Eradication Initiative, a partnership run by the World Health Organization. The program is plainly paying for something more than the vaccine.[27]

Could it be lack of infrastructure? Vaccine delivery depends on the "cold chain," the transport of vaccines to individuals under steady refrigeration. It is much harder to provide a good cold chain in environments with poor roads, old trucks, unreliable power, and little access to chilled storage. Yet the nations of the world managed to eradicate smallpox at a time when asphalt, automobiles, and refrigerators were even rarer than they are today.[28]

Of course, technology, money, and infrastructure are needed. Yet, so often, those things are present, and the desired impact still fails to materialize. It's no coincidence that countries whose health-care systems are in chaos see more than their share of avoidable illness. The fact that vaccine-preventable diseases still plague the developing world gives the lie to anyone whose faith is placed in packaged interventions in and of themselves.

Humanity Not Included

Though we look to packaged interventions for large-scale impact, the interventions themselves depend on something else. They require a substrate of positive intent and high capacity among individuals and institutions–the exact substrate that is in stunted supply where social challenges persist.

Who forms the critical human substrate? In my team's educational technology projects, we saw that the key people were the researchers who developed the technology, the teachers who applied the technology, and the students who used the technology. These three groups have analogues in other contexts as leaders, implementers, and beneficiaries.

Leaders are those who have power over a packaged intervention–what form it takes, whether it is applied, and how. They may be nominal leaders, such as government ministers or nonprofit chief executives, or they may be the people who create technologies, devise policies, supply funds, or otherwise influence the design of an intervention.

Few dispute the critical role of leadership. In microcredit, for example, it matters greatly whether those in power–microfinance institution heads, policymakers, investors, and so on–are committed to supporting low-income borrowers. If they allow personal greed–or even an overeager intention to serve more people–to supersede that commitment, the impact will sour. As Yunus noted, when growth became the chief goal of microcredit, "banks needed to raise interest rates and engage in aggressive marketing and loan collection. The kind of empathy that had once been shown toward borrowers when the lenders were nonprofits disappeared."[29]

Unfortunately, empathy cannot be packaged, and it doesn't increase with the size of an intervention. As much as Yunus rails against profit-making microfinance organizations, their rise continues. Compartamos is growing even as it turns a blind eye to the plight of former borrowers. Stories like that of Hernández, the sock-knitting entrepreneur, leave Carlos Danel, Compartamos cofounder and executive vice president, unmoved. "A lot of people have suggested that financial inclusion can be a poverty alleviation tool," he said. "We're not out to prove that."[30]

Implementers are the second vital group. Consider again microcredit, whose methodology leaves little room for error. It's not easy to make small loans to poor borrowers. They have no formal collateral. Borrowers must be organized into groups that hold regular meetings. Loan officers must travel to remote villages, often carrying wads of cash that tempt thieves. Lending decisions must be made on the basis of scant formal documentation. Banks and other sources of capital must be courted. Local law and customs must be obeyed. And all of that activity must break even financially. In fact, microcredit has a niche precisely because this process is too costly for formal banks; they can't work directly with very poor borrowers without losing money. When it works, microcredit is a marvel of implementation that combines a high-precision process with compassion for borrowers.

Implementers are the individuals and institutions who execute a packaged intervention. They install, operate, and maintain technology. They build, run, and manage institutions. They adapt, announce, and enforce policies. They set up, control, and administer systems. But, as essential as they are, implementers are rarely appreciated. Recognition goes to individual leaders such as Yunus rather than to frontline Grameen Bank employees. Or to Edward Jenner, who identified the smallpox vaccine, rather than to the health workers who administered the doses. Or to Wael Ghonim, who started a Facebook page, rather than to the nameless protesters in Tahrir Square. We could hardly tell the stories of interventions if we had to name every responsible party, but by overlooking implementers, we take good implementation for granted. It's another quality that can't be packaged.

Beneficiaries form the third group whose willing and able participation is required. But leaders and implementers often overestimate the desire and capability of beneficiaries. Yunus, for example, says of poor households, "All they need is financial capital."[31] Perhaps out of respect, he dismisses any need to train them: "The fact that the poor are alive is clear proof of their ability. They do not need us to teach them how to survive." He goes as far as to say that training programs are "counterproductive."[32] Yet several studies hint that no more than a quarter of households eligible for microloans actually take them on–poor people themselves know that they need something more.[33] Abhijit Banerjee and his colleagues have found that it's mostly experienced entrepreneurs who apply loans to business expansion.[34] And Vijay Mahajan, chief executive of the Indian microfinance institution Basix, cautions that loans can "do more harm than good to the poorest."[35]

People get used to chronic injustices. And impoverished people spend their energies dealing with the travails of daily life. So, just like with anyone else, motivation and ability can't be taken for granted; they often need to be built up. Everyone wants to live a long life, but couch potatoes struggle to keep up exercise regimens. Most people value education, but few can sustain hours of directed study day after day, year after year, on their own. Everybody wants freedom, but not everyone will risk their own lives on the frontline of protest. Even vaccines, which more or less work upon injection, have no effect if potential beneficiaries don't want them. In some parts of the developing world, people fear that vaccination campaigns are attempts to sterilize them. And in some parts of the developed world, people resist vaccines out of misguided fear that they cause autism and other problems.[36]

The Curse of Packaged Interventions

Leaders, implementers, and beneficiaries. This trinity determines packaged-intervention success, but their all-important traits can't be packaged. No technology includes the empathy and discernment needed in leaders. No law bundles capable implementation. No system guarantees that beneficiaries will want what others believe is good for them. Exactly that which makes a packaged intervention work can't be mass-produced.

Technologists and technocrats might hear a challenge in that statement. They hate to admit the existence of systemic obstacles that can't be overcome by sheer brilliance. I'm often approached by young hacktivists who ask me for feedback on their projects. They start by describing, say, a mobile SMS text-messaging system that would mass deliver health advice to poor families who don't have access to medical services. I explain to them that poor people are no different from you or me. They have little tolerance for spam. Assuming that they can read at all, they'll probably ignore messages from an unknown source. Why would they take unsolicited advice over that of their friends, relatives, or the village priest? Any project like that works only when it piggybacks on some established form of trust.

Similarly, any solution–technological or otherwise–that improves education depends crucially on parents, students, teachers, and principals. Anything that augments agriculture depends on suppliers, farmers, extension officers, and produce buyers. Anything that enhances governance depends on bureaucrats, administrators, leaders, and citizens. The lack of attention to these necessary human factors leads to the broken medical equipment, shuttered offices, and collapsed democracies that litter the history of social causes.[37]

It's no exaggeration to say that packaged interventions have a kind of curse. Ironically, it is exactly where there are extreme social problems that the human intent and capacity required to make good on packaged interventions is missing. It's the subpar schools we most want to fix with computers that lack teachers, principals, or IT staff who can make good use of technology. It's the dysfunctional governments we most want to replace through elections that lack the institutions, civil society, and armed forces willing to hold up democracy. And it's the jagged social fissures that we most want to stitch up with laws that lack the interpersonal trust and mutual respect needed for healing.

The Iron Law of Social Programs

Peter Henry Rossi was among the most influential sociologists of the mid-twentieth century. He was a frequent presence in Congress, where his testimony influenced US domestic policy. Colleagues described him as devoted to social causes and ardent about hard evidence. He once stared down a colleague in a public debate and said, "My data are better than yours."[38]

After a career spent studying programs intended to combat crime, poverty, and homelessness, he published an unusual paper in 1987 called "The Iron Law of Evaluation and Other Metallic Rules."[39] In it, Rossi made a startling claim. The Iron Law of the title stated, "The expected value of any net impact assessment of any large scale social program is zero." That is, on average, large social programs show no impact.

Rossi spent much of the paper considering this discouraging conclusion, and his analysis foretold the problems of packaged interventions. He explicitly noted three reasons why social programs so often fail when they go big.[40]

Rossi's first two reasons related to bad program design. Either there is a problem with the theory of the intervention, or there is a problem with how the theory is put into practice. These issues correspond to problems of leadership: Those with influence over the packaged intervention have gotten something wrong.

Rossi's third reason was faulty implementation. One sentence in Rossi's essay pinpoints what some people today call "pilotitis," in which social programs seem to do well in pilots, but fail at large scale: "There is a big difference," wrote Rossi, "between running a program on a small scale with highly skilled and very devoted personnel and running a program with the lesser skilled and less devoted personnel that YOAA ordinarily has at its disposal." (YOAA is Rossi's own coinage; it stands for "your ordinary American agency.") In other words, pilots often succeed because of the immense talent and motivation brought to bear on a new program. Once the program is handed off to an indifferent bureaucracy, though, it fails. The secret sauce is not in the program specs, but in the implementers.

All That Glitters

What isn't a packaged intervention? The caring attention of a good teacher. Citizen participation in a protest march. The capable execution of a vaccination program by a well-managed health-care system. A political leader's decision to do what's right in spite of pressure from special interests. Human virtues can't be packaged.

But if packaged interventions don't include the critical human components of social change, then isn't it tautological to say that packaged interventions aren't enough? It does seem pretty obvious. But if it's obvious, experts of every stripe nevertheless treat packaged interventions as if they were complete solutions. Critical decisions in public policy are made on the basis of this or that packaged intervention. Academics write papers arguing one packaged intervention's superiority over another. Foundations allocate budget items for specific packaged interventions. Journalists glorify new packaged interventions. And activists push packaged interventions like drug dealers pushing narcotics.

Why do people do this? Many of the reasons are less than noble: Some seek profit. Others, fame. And still others just want to win a proxy war of egos: My program improves a gazillion lives; how about yours?

But even the most sincere, devoted people insist on the glittery magic of their packaged interventions. Take Yunus, a man who cannot but be admired for his lifelong dedication to fighting poverty. He claims that loans are the only thing poor people need: "Giving the poor access to credit allows them to immediately put into practice the skills they already know."[41] He dismisses concerns about whether their skills are sufficient. "Not one single Grameen borrower requires any special training," he says. "All they need is financial capital."[42]

Yunus isn't the only one. Here is John Hatch, founder of the nonprofit Foundation for International Community Assistance, describing his philosophy of microcredit: "Give poor communities the opportunity, and then get out of the way!"[43] Kiva.org, an online portal where individuals can contribute to loan capital, asserts, "Low-income individuals are capable of lifting themselves out of poverty if given access to financial services."[44] Opportunity International, another large microcredit organization, says simply that microcredit is "a solution to global poverty."[45]

The people making these claims, though, contradict their public statements in their own work. When I met Yunus, he spoke of the Herculean efforts required to grow Grameen Bank: nurturing capable borrowers into employees, finding new ways for women to earn income, spurring house-bound wives to enter public life, and on and on. It was clear he had climbed mountains beyond mountains, all the while holding the hands of his borrowers and pulling them along. Even as he denies that poor people need training, he describes a unique culture at Grameen Bank that all of his borrowers join. At each meeting, for example, they recite "Sixteen Decisions," ranging from, "We shall build and use pit latrines," to, "If anyone is in difficulty, we shall all help him or her."[46] Yunus may not think of this as training, but it's definitely a lot more than just credit. You can be sure that the folks at Compartamos aren't reciting daily oaths to help their borrowers in times of difficulty.

Yunus is clearly aware of the complexities. Still, something causes him to market an oversimplified partial solution. In this, he is just like promoters of digital technology. Recall Ghonim: "If you want to liberate a society, just give them the Internet."

These paeans to packaged interventions influence budgets for social causes. Global spending on interventions is hard to track, but even the rough estimates are staggering. According to the Microfinance Information Exchange, which collects financial data from microcredit organizations around the world, in 2012, 1,161 participating microcredit institutions served over 92 million borrowers in low-and middle-income countries with a total loan portfolio of $94 billion.[47] That's a low estimate. Several US agencies have committed $125 million to the Global Alliance for Clean Cookstoves, which encourages the world's poorest households to exchange their makeshift stoves for less polluting alternatives. A bipartisan bill introduced in the Senate backs the effort. In 2009, Richard Heeks, a professor at the University of Manchester and a pioneer in the study of information technology for international development, cited estimates of IT spending on the developing world ranging from $2 billion to $840 billion, depending on what was counted.[48] He concluded that "hundreds of millions of US dollars per year" were devoted to one-off IT projects, and that "tens of billions of US dollars per year" went to IT infrastructure specifically for the developing world. These are large expenditures considering that in 2012, the world's total foreign aid directed to education was just $12 billion.[49] Packaged interventions rule formal efforts to effect social change.

But, as we've seen, rigorous studies of both microcredit and PCs in schools show that neither is effective on its own. Widespread use of the Internet isn't ending poverty or inequality, even in wealthy America. Decades of cookstove design have only produced ineffective or underused models.[50] Studies of charter schools show little improvement, on average, over public schools, especially when effects on students from similar backgrounds are compared.[51] All of this is to repeat that packaged interventions in and of themselves don't guarantee positive social change, whatever "opportunities" they may provide.

Nevertheless, once an intervention is wrapped up in a tidy package with a bow on top, it's hawked as *the* special sauce, *the* silver bullet, *the* alpha and omega. Credit would be an end to poverty if borrowing created jobs and taught people employable skills. Negroponte's vaccine analogy would be apt if a moment with a laptop ensured years of education. Egypt would be an oasis of democracy if an election were enough to cause good governance.

Of course, technologies *can* enrich lives; voting *can* empower citizens; and microcredit *can* lead to better livelihoods. But "can" is not always "will." Modern society fetishizes technocratic devices, but it's a human finger on the on-switch and a human hand at the controls. Why are we so enamored of shrink-wrapped quick fixes? Why do even those of us who know better tout them as real solutions? The reasons run deep and have been centuries in the making.

CHAPTER 5

Technocratic Orthodoxy

The Pervasive Biases of Modern Do-Gooding

In 1987 I moved to Cambridge, Massachusetts, to enroll as a freshman at Harvard. Bacchanalian parties weren't really my thing, so I had a lot of time on weekend evenings. It felt a little lonely in my dorm room while loud music thumped its ways through the walls, so I took walks around the city and found that it catered to a vice of my own: books.

At the time there were as many as thirty booksellers in the few blocks that made up Harvard Square. Local tour guides boasted that there were more bookstores per square mile than anywhere else on earth. I remember marinating in the musty smell of McIntyre & Moore, seeking enlightenment at the Thomas More Bookshop, skimming leatherbound copies of Plato at Mandrake, indulging guilty pleasures at Science Fantasy Books, and bottom-feeding at Buck-a-Book. My favorite was WordsWorth on Brattle Street. I spent hours upon hours in its two floor-to-ceiling stories of glorious inventory, all sold at discount.

Today those shops are gone. With no more than seven or eight bookstores left, Harvard Square has lost its bibliophilic bragging rights. Or, possibly worse, with those seven or eight shops, it might still be the Bookstore Capital of the World. Everywhere you look, brick-and-mortar bookstores are being steamrolled under an online juggernaut.

Amazon sold its first book online in July 1995. By 2013 the company captured as much as a third of all US book sales and 60 percent of all e-books.[1] *Publishers Weekly*, "the bible of the book business," might as well be called *Amazon Watch*. Between the best seller lists and the book reviews, the articles seem to fall into two categories: shock and breast-beating about Amazon's latest ploys, and the forced cheer of publishers, librarians, and retailers frantically seeking the bright side of a cataclysm. The collective fear is that the industry will only have room for Amazon, its customers, and a few best-selling authors. Midlist writers and discerning publishers will cease to exist. The infinite variety of genuine literature will be reduced to *Fifty Shades of Grey*. Main Street will never again be a site for serendipitous browsing among stacks of lovingly selected books.

Meanwhile, this angst goes largely unnoticed by consumers. They might remember the neighborhood bookstore with nostalgia, but they're happy that they can buy new e-books for $9.99 and have access to just about any text at their fingertips. Is there really a crisis in publishing? Maybe old-school publishers just can't learn new tricks.

Whatever you think of Amazon, you're likely to assume that digital caused these changes. Here's a twenty-years-young Internet-only bookseller that almost singlehandedly built the e-book market–what other explanation could there be?

History points the finger elsewhere. The book business has been cutting costs and courting best sellers for decades, leading to plenty of anxiety along the way. After Penguin began mass-marketing the paperback in 1935, George Orwell wrote, "In my capacity as reader I applaud the Penguin Books; in my capacity as writer I pronounce them anathema."[2] Starting in the 1970s, it was Barnes & Noble's turn to lay waste to the bookselling establishment. With its block-spanning superstores, aggressive discounts, and mail-order sales, the brand grew mighty, eventually acquiring and killing mall mainstays B. Dalton and Waldenbooks, which themselves had stolen business from independents.[3] And publishers have been chasing the blockbuster since long before the Internet. In *Merchants of Culture: The Publishing Business in the Twenty-First Century*, sociologist John B. Thompson highlights an industry turn toward fewer titles, with each expected to have bigger sales.[4] He traces this trend back to the 1960s, when rapid consolidation of publishing houses began. The mega-merger of Random House and Penguin took place in 2013, but Random House itself is an amalgamation. It swallowed up Ballantine, Bantam, Crown, Dell, Doubleday, Fawcett, Fodor's, Knopf, and so on down the alphabet.

In short, Amazon and its digital ways are an extrapolation–an amplification–of pre-digital patterns. On the one hand, more books of greater variety are being published. There is a growing population of writers, and publishing is being commoditized.[5] On the other hand, the books that receive widespread attention and land on best seller lists are a dwindling proportion of the total. The first trend is sometimes called "the long tail"; the second represents a "winner-take-all" economy.[6] Commentators tend to highlight one or the other of these phenomena, but both are happening at once. Together they cause the shrinking middle that is the hallmark of any industry–music, movies, manufacturing–whose product can be replicated and distributed cheaply. Indeed, in the book business it is widely acknowledged that fewer and fewer authors are able to make a living through writing. The squeeze, however, long predates digital tools.

Amazon's case reinforces a consequence of the Law of Amplification: Technology trends aren't always *technology* trends. They often have a pre-digital history. Amazon is an extension of a larger book industry driving toward cold efficiency at the expense of less tangible merits. Mix in digital technology, and something like Amazon is sure to come along. (And it would come again, if Amazon were to perish.)

To understand Amazon properly, it's not enough to know its technology tactics. You also have to know the history of paperbacks, the expansive aggression of Barnes & Noble, and the inclinations of print publishers, which, even as Amazon threatens them, are cut from the same cloth. You have to look at the non-digital context.

Similarly, to better understand our technology fixation, it's important to recognize its larger social and historical context. As I began to doubt the hype around packaged interventions, I wanted to see if I could bypass their problems. Maybe there were other approaches to social change. So I engaged with three ideas that have growing support–randomized controlled trials, social enterprises, and happiness as a goal. These are largely unrelated efforts, but they all have great merit and are well-regarded within their specializations. Promisingly, each had a potential claim to exorcising the curse of packaged interventions.

The Randomista Revolution

In July 2011 I visited a school in Kotra, a small village in southern Rajasthan. The small hut had white plaster walls and a thatched roof. About twenty children wearing bright blue uniforms sat on the floor in two circles, one led by a male teacher while the other quietly worked arithmetic problems. I kneeled next to a couple of the students and watched as they marked small slates of blackboard with chalk. They were practicing subtraction with three-digit numbers. A couple of students stumbled over borrowing a one from the next digit, but most of them worked the arithmetic correctly. They were focused, and the teacher was attentive.

The school was operated by a nonprofit called Seva Mandir that has nurtured rural communities in two districts of Rajasthan for over forty years. I was there to visit a project I had read about in a research paper.[7] The paper's first author had an innovative idea to combat teacher absenteeism, and it was tested in Seva Mandir's schools: Teachers would take digital photographs of themselves and their classes at the beginning and end of each day they were at school. Then, the teachers' pay would be linked to the number of days each month for which they supplied photos. To prevent teachers from cheating, the researchers devised special tamper-proof digital cameras.

It was an interesting idea, but what brought me to Rajasthan that day was not the use of digital technology per se. I had already seen hundreds of technology projects by then. What made this project unique was that world-renowned researchers had used a rigorous methodology to establish something that seemed to contradict the Law of Amplification. The research team was led by Esther Duflo, a brilliant MIT economist who counts among her honors a MacArthur "genius grant" as well as the John Bates Clark Medal, a good predictor of future Nobel laureates. As a pioneering member of the Abdul Lateef Jameel Poverty Action Lab (JPAL), Duflo has been a tireless advocate for the use of randomized controlled trials (RCTs) to verify the value of antipoverty programs. This is the methodology used in clinical medicine, whereby a control group establishes a baseline against which the effectiveness of a treatment can be compared. In applying the rigor of hard science to social questions, Duflo and her colleagues are revolutionaries. Rivals and supporters have nicknamed them "randomistas."

In a paper describing the effort, Duflo and her colleagues reported dramatic results. As expected, attendance was captured by the cameras, and teachers showed up more often to class. And the teachers' consistent presence encouraged students to show up, too. Two photos a day kept absenteeism at bay.

But reducing absences isn't the end goal–better education is. And here the experiment had an even more impressive outcome. Compared with students in a control group, the students in the camera-monitored classrooms performed better on tests of math, reading, and writing. The teachers who bothered to show up for the photos must have stayed to teach, and, apparently, their teaching was effective. This finding is striking, because other studies of developing-world schools–including one by Duflo herself–show that time sitting in classrooms doesn't always translate to significant learning.[8]

For me, Kotra offered new hope. Maybe in some circumstances, technology could work wonders on its own. In this case, the cameras seemed to cause strong educational outcomes with no special attention paid to teachers. Maybe I was premature in declaring the futility of isolated packaged interventions. I was curious about the tamper-proof cameras and what it was about the project as a whole that made it work. Could its lessons be generalized to other contexts?

Packing Tape, But No Packaged Teaching

After the class in Kotra, I asked the teacher to show me how he took the class photos. He took a camera out of a cabinet and flipped through photos on the display. Each showed a teacher standing at attention next to three rows of uniformed students. As the teacher handed the camera to me, I was a bit disappointed to find that the "tamper-proof camera" turned out to be unexpectedly low-tech. They were inexpensive Yashica digital cameras with cellophane packing tape placed over the controls.[9] Still, the pristine condition of the tape suggested that the tamper-proofing achieved its purpose.

Something else I learned on that visit, though, caused far greater disappointment. The authors wrote, "Our results suggest that providing incentives for attendance in nonformal schools can increase learning levels."[10] The paper was sure to please anyone who believes in meritocratic rewards for individual effort. It was titled "Incentives Work," and it attributed all the educational gains to the differential pay provided to teachers for photo-proven attendance. If you had only the paper to go by, you'd believe that wherever there is high teacher absenteeism, incentive pay for photo-validated attendance would be sufficient to improve learning. As JPAL's website puts it, "If teacher attendance can be improved this should flow through into improved test scores."[11]

But this finding may not stand up in even in Kotra, much less in low-income villages elsewhere. The researchers neglected the full context of the project. After visiting the school, I spent a few hours with Seva Mandir staff at their offices in Udaipur. By then, I had seen many of their other initiatives, and it was clear that Seva Mandir was a devoted, well-run organization that worked small miracles in almost all of its programs. I wondered if what I saw with the camera-monitoring program wasn't so much a counterexample to the Law of Amplification as yet another instance of it. Maybe the results weren't caused by the camera program alone, but by the camera program as implemented by a strong organization committed to quality education.[12]

Conversations with the staff confirmed my hunch. When Seva Mandir began working with schools in Kotra several years before the study, the teachers were inexperienced and their lessons poor. In keeping with the pace of rural life, parents didn't care much about either teacher or student absenteeism. For years, Seva Mandir worked extensively with the schools, teachers, and parents to improve pedagogy and to persuade everyone involved of the value of daily attendance.

By the time camera monitoring was put in place, the teachers were more motivated and effective than when they had started. They were operating with a level of commitment and ability rare in rural India. The teacher I met in Kotra, for instance, was far more skilled than others I had seen in government schools. His lesson was pitched well, and he encouraged the right kind of interaction.[13] So, while camera-monitored, attendance-based pay surely had its impact, the educational benefits came on top of years of effort by Seva Mandir and the Kotra community.

None of this is mentioned in the paper. There is only one paragraph about the groundwork that Seva Mandir had laid prior to the trial, and it focused narrowly on the efforts surrounding teacher attendance, not on the other painstaking labors to improve teacher capacity or community expectations. As a result, the paper leaves the impression that camera monitoring is the only thing you need to implement in order to improve learning in schools. In effect the study attributes the nutritional value of a whole meal to dessert alone.[14]

Originally, I saw in RCTs the hope of transcending the limitations of packaged interventions. Duflo and other randomistas discourage broad theories to guide social change. Instead they encourage lots of experiments to see what works.[15] It's a sensible, data-driven approach, and, as a result, RCTs are seeing a surge of popularity. Not only economists but political scientists and sociologists are now running more RCTs.

Critics have noted, however, that not everything can be tested with an RCT.[16] Some questions are inherently harder to explore with the technique. In practice, certain kinds of questions are systematically neglected, and some programs never receive the imprimatur of an RCT. The methodology excels at comparing packaged interventions against one another. If you want to know whether camera-based monitoring or a midday meal is more likely to help students do better on tests, an RCT is great. But comparing packaged interventions against non-packageable alternatives is much more difficult.

Here's why. In order to run an effective RCT, researchers must carefully ensure a number of conditions. The experimental group must be selected so as to be without bias (hence, randomization). The control group must remain unaffected by the experiment. The intervention must be implemented exactly as defined. Data must be collected. And all of this must be done to exacting standards. To guarantee these requirements, researchers prefer to run RCTs in partnership with capable organizations such as Seva Mandir, that is, organizations capable of following onerous instructions. But by working with competent implementers, the experiments necessarily occur in special conditions that aren't common elsewhere.

When the results of an experiment can't be generalized beyond their immediate context, scientists call it a problem of "external validity." Duflo and other randomistas say external validity can come by repeating studies in different contexts. Technically, that's true, but it's not easy to do for testing the value of capable implementation. The very thing you need for a good controlled trial would have to be suspended for a part of the experiment.[17] This is the RCT's Achilles' heel.

You can hear reverberations of Rossi's Iron Law of Evaluation. There is, he wrote, a world of difference "between running a program on a small scale with highly skilled and very devoted personnel and running a program with the lesser skilled and less devoted personnel."[18] The Law of Amplification is also in effect: Packaged interventions work in proportion to the capacity brought to bear. If you conduct experiments entirely in air, where institutional capacity can be taken for granted, then you don't know how the intervention would run under water, where capacity might not be present.

To be clear, I'm not against RCTs in and of themselves. I have applied RCT methodology in my own research. I believe that packaged interventions should be verified by RCTs more often. I'm on the board of JPAL's sister organization, Innovations for Poverty Action, a nonprofit that also runs RCTs and whose work I deeply respect and support.

But if RCTs are an essential tool of decision-making for social policy, they still need to be placed within a larger conceptual framework. If a world-renowned economist and careful experimentalist such as Duflo can't avoid the pathologies of packaged interventions, it's not clear that other researchers running RCTs can either. A single methodology cannot be the sole paradigm for determining what's right for social change. The problem is not RCTs themselves as much as careless interpretation of their results.[19] An RCT is just one good tool in the toolbox of program evaluation. Good tools are important, but it's even more important that architects and artisans use the right combination of tools in the right way for each decision-making task.

Eradicating Poverty Through Profits?

Another fashionable trend sees practitioners applying for-profit business approaches to social causes. The idea was put forth seductively by C. K. Prahalad, a professor of business at the University of Michigan. In his 2004 book *The Fortune at the Bottom of the Pyramid*, he wrote that the 4 billion people in the world who live on less than $2 a day could be enriched if they were viewed as a business opportunity.[20] Prahalad's motto is captured in the book's subtitle: *Eradicating Poverty Through Profits*.

According to Prahalad, governments and nonprofits have been going about things all wrong, especially when it comes to poverty. They commit two sins hateful to any business: First, they don't cover their own costs; second, they're unable to reach large numbers of people. "Charity might feel good," Prahalad observed, "but it rarely solves the problem in a scalable and sustainable fashion."[21]

Prahalad's proposal was to sell low-cost products and services with the goal of fostering poor people's "capacity to consume." These people, he said, were "value-conscious consumers," and affordable consumption opportunities served them by "offering them choices and encouraging self-esteem."[22] Prahalad's rhetoric rehashed the themes of mainstream economics: individual choice, market freedom, and poverty alleviation without politics. He promised that there were "sustainable win-win scenarios where the poor are actively engaged and, at the same time, the companies providing products and services to them are profitable."[23]

If this sounds too good to be true, that's because, by and large, it is. Although Prahalad's book is crammed with case studies, none of the examples actually shows a firm increasing its profit margins while improving the lives of poor customers. Aneel Karnani, another University of Michigan professor, took a critical look at Prahalad's nine case studies and found that four were companies catering primarily to middle-income consumers, two were nonprofits sustained in part by public funding, and two were projects that hadn't seen any profit. Only one was a for-profit corporation that made money selling to poor consumers.[24]

That last case is worth examining, because, as much as it seems to fit Prahalad's thesis, it still comes up short. The company in question was the Unilever subsidiary Hindustan Lever Limited (HLL), and the product was soap, possibly Prahalad's most widely cited example. Diarrhea is deadly for many children in India, but it can often be prevented by hand washing. So HLL sold Lifebuoy soap in small sachets that were affordably priced for "bottom of the pyramid" consumers.

It's one thing to sell soap, though, and it's another to instill hand-washing habits in people who've never learned the germ theory of disease. HLL tried two approaches. It ran a marketing campaign designed by the Madison Avenue giant Ogilvy & Mather, and it established a public-private partnership with the Indian government. Together, HLL and the state ran a multimillion-dollar, multi-platform campaign to convince people to wash their hands. Which worked better, the corporate marketing scheme or the public-private one? By Prahalad's own admission, "although scalability seems to be greater with the [public-private partnership], benefits to corporate sales lie with" paid marketing efforts. In other words, while profits were best served by a corporate campaign, greater health impact came by way of taxpayer-funded social marketing.[25]

So none of the nine cases Prahalad held out as paragons of "eradicating poverty with profits" were any such thing. Either they didn't serve the poorest people, or they didn't make much of a profit. This shouldn't come as a surprise. We already know how hard it is for the private sector to deliver services to poor populations on its own. For instance, despite many experiments by determined entrepreneurs, there isn't a system in the world that delivers universal health care or universal education without at least some support from the public sector. Private efforts either fail to generate enough revenue to keep going, or have to swim upstream to richer waters.

What's most twisted about Prahalad's logic is the suggestion that poor people can somehow consume their way out of poverty. As if their poverty were a function of what they can buy, not what they can earn. Karnani more sensibly argued that the best way to support poor people was to help them become higher-income producers, not consumers.[26] Well-paid employment makes people richer. Godfather of microfinance Muhammad Yunus has come to a similar conclusion and urges social activists to start what he calls "social businesses."[27] They differ from Prahalad in seeking to employ low-income workers, not just sell to them.

Asocial Enterprises

Prahalad has had some influence in corporate circles, but it's a variation of his idea called "social enterprise" that has really taken off. In business schools, engineering departments, and venture capital firms, social enterprises–start-up businesses that try to serve a social good through a viable business–are all the rage. Social entrepreneurs model themselves on the Steve Jobses and Mark Zuckerbergs of the world, not realizing that successful businesses are successful because they have carefully chosen their customers, not because they have a foolproof Midas touch. Apple is a profitable company not just because it designs superior products but also because it chooses the world's wealthiest people as its market. It would hardly survive if it were constrained to selling $400 iPhones to individuals who earn less than that in a year. Even ad-based business models only work if advertisers are willing to pay for ads, which they won't do if the eyeballs belong to people who lack disposable cash.

Toms Shoes has found a way around this problem. Its founder, Blake Mycoskie, is considered a social-enterprise pioneer. Toms is a for-profit company whose marketing strategy–touted in boldface at the top of its website–is "one for one": For every pair of shoes sold, Toms donates a pair to a "person in need."[28] The firm is wildly successful.

Since its founding in 2006, it has handed out over 10 million shoes, which implies cumulative revenue of about half a billion dollars. It has recently expanded into eyeglasses with the same one-for-one promise. Bill Clinton once introduced Mycoskie–whose rakish good looks recall Jim Morrison–as "one of the most interesting entrepreneurs [I've] ever met."[29]

Despite its performance, Toms has also come under scathing criticism. Some say that giving away shoes to poor communities stunts local economies and perpetuates a culture of dependence. Others note that Toms could redirect its giving to something more lasting than straw-and-canvas shoes. Even if their gifts cost just $5 each to make and transport, that's still $50 million that could have been better spent. Still others find Toms to be little more than a sweatshop, exploiting cheap Chinese labor and photos of barefoot children for hefty profits.[30]

There is some truth to these points, but the real problem is deeper and subtler. I suspect Mycoskie's critics wouldn't be mollified that he has responded to them by opening factories in countries where he gives away shoes.[31] There is something more insidious at work.

Though Toms trumpets its commitment to addressing "hardships faced by children growing up without shoes," it's not very open about what customers are actually paying for. Toms doesn't disclose its financial statements. What we do know is this: Thanks to a recent agreement to sell 50 percent of his stake to Bain Capital, Mycoskie stands to gain as much as $300 million on top of whatever he's paid himself so far as sole owner and CEO.[32]

Imagine if you made a $50 donation to a nonprofit, and out of that, say, $10 went to the cause while $10 came back to you as a thank-you gift and $30 facilitated the executive director's multimillion-dollar bonus. You wouldn't stand for it. Yet this is effectively the Toms "business model." And from it, the founder not only richly profits but then is weirdly hailed as a social activist's hero. Something doesn't seem quite right, and a comment made by a principal at Bain Capital–the same company that cast a dark shadow on Mitt Romney's presidential campaign–provides a clue why: "We believe that the one-to-one promise is fundamental to the brand."[33] Or, to put it another way, Toms is successful because of a cynical marketing pitch catering to slacktivist consumers who equate generosity with purchasing two cheaply made pairs of shoes and giving one away.

In the end, Toms is a shoe company with a social responsibility arm. Mycoskie deserves credit for being a shrewd entrepreneur and spending some of his growing fortune on charity. Otherwise, though, Toms is not that different from Nike: They both sell overpriced shoes to brand-conscious customers, exploit cheap developing-world labor to pay their executives well, and spend a portion of their revenue on charitable causes (in Nike's case, the nonprofit Nike Foundation).

But Toms does one more thing: By misleadingly presenting itself as primarily interested in charity, the company diverts the goodwill of people who might otherwise engage more deeply in a cause. In what psychologists call moral self-licensing, people use past good deeds–even minor ones–to excuse future apathy.[34] So there's a good chance that many Toms customers skimp on more worthwhile efforts, something they probably wouldn't do if they bought their shoes from Nike, which runs its own social responsibility initiative but with less self-congratulatory fanfare. The greater jeopardy, though, is a broader *societal* self-licensing: By playing up efforts like Toms, we as a society fool ourselves into believing that the world's problems can be solved by enlightened consumerism.

That's the problem with social-enterprise hyperbole in general. Its noisy buzz draws attention away from effective government and nonprofit approaches to social causes.[35] Large aid agencies such as United States Agency for International Development (USAID) and the Ford Foundation have begun spending their precious funds on for-profit entities. They hope that their one-time donations will turn into everlasting economic engines that also churn out public goods. As Jonathan Franzen observed in his novel *The Corrections*, "The more patently satirical the promises, the lustier the influx of American capital."[36]

It's just not true, though, that not-for-profit models can't have sustainable impact at scale: Most developed countries have government-subsidized universal health care.[37] Even Rwanda has national health insurance that covered 92 percent of its population in 2010.[38] In

2011, 91 percent of the world's school-aged children were enrolled in primary education, most of them in government-subsidized schools.[39] The International Committee of the Red Cross has helped "victims of war and armed violence" around the world for over 150 years. It raises more than $1 billion a year, almost entirely through private donations.[40] Other nonprofits, such as CARE, Goodwill, Human Rights Watch, Oxfam, and United Way, are similarly large and effective. However much governments and nonprofits are criticized for their waste, none of them generates $300 million payouts for their chief executives and pretends that's an effective way to direct goodwill.

I've participated in several social enterprise competitions as a coach, mentor, or judge. Talented twenty-somethings feverishly pitch projects, hoping, like Blake Mycoskie, to "do well by doing good." The projects are tested mainly for their financial sustainability (read "profitability"), their scalability (read "market penetration"), and novelty and uniqueness (read "potential monopoly power"). In the mad rush to conjure money out of hoi polloi, the "social return on investment" often becomes an afterthought. It's as if there's no point to saving lives or teaching children if you have to keep paying to do so.

Real social change is no easier to achieve with social enterprises than with not-for-profit models. The hype, though, allows business success to be confused with social impact.

Happiness and Its Discontents

Social causes seek economic prosperity, social justice, human dignity, and expanded freedoms, so packaged interventions aim for these goals, too. But what if the goals are themselves misguided?

In 1972, King Jigme Singye Wangchuk of Bhutan proposed an alternative measure of progress. He announced that instead of Gross National Product, his country would judge itself by what he called Gross National Happiness. And before we deride a young king of a small, far-off land for his idealism, it's worth remembering that Thomas Jefferson, representing a once young, once far-off land, enshrined "the pursuit of happiness" as an inalienable right on par with life and liberty.

Jefferson and the Bhutanese king knew what they were talking about. Philosophers have proposed happiness as the highest good and the ultimate goal of human activity at least since the Buddha and Aristotle. A couple of thousand years later, Jeremy Bentham and John Stuart Mill expanded the notion to whole societies. In their utilitarian philosophy, the goal is the greatest good for the greatest number.

Amazingly, in the past decade or so even no-nonsense economists have started taking happiness seriously. Neuroscientists such as Richard Davidson have shown that certain kinds of brain activity–measurable by functional magnetic resonance imaging (fMRI)–are correlated with self-reports of happiness.[41] Alluding to this work, the eminent British economist Richard Layard wrote, "Now we know that what people say about how they feel corresponds closely to the actual levels of activity in different parts of the brain, which can be measured in standard scientific ways." Satisfied that happiness was real, Layard wrote an entire book arguing that happiness, not wealth, should be the basis for public policy.[42]

All of this adds to an ongoing case made by scholars, policymakers, and activists who argue that today's dominant metrics of national progress are deficient. In 1995, *The Atlantic* asked, "If the GDP Is Up, Why Is America Down?"[43] The country collectively repeated that question in 2009. GDP recovered from the recession, but employment didn't. A good metric should correlate with the overall well-being of a country. What's the point of a metric that increases while so many people are miserable?

King Wangchuk was ahead of his time. There's no point to wealth or social change unless they lead to greater happiness. War, illness, hunger, thirst, poverty, oppression, ignorance, unemployment, and powerlessness are problems in great part because they're obstacles to happiness. Prosperity leads to the material requirements for happiness. Justice seeks the moral conditions for happiness. Dignity lays the material and political basis for happiness. And freedom, as Nobel-laureate economist Amartya Sen wrote, allows people to live "lives we have reason to value"–that is, lives we think would make us happy.[44] If there were a community of poor, ignorant, marginalized people who were nevertheless always happy, we wouldn't feel any need to help them. In any case, they probably wouldn't want our "help."

With this recognition has come a wave of public interest in private happiness. Books with titles such as *Authentic Happiness*, *Stumbling on Happiness*, *The Happiness Hypothesis*, and *The Happiness Project* are proliferating, competing to counsel us on how we can individually be happier.[45] Global leaders have also turned to happiness and related concepts. In 2009, French president Nicholas Sarkozy commissioned a group featuring five Nobel Prize winners to devise a metric that captured true quality of life.[46] In 2010, British prime minister David Cameron prompted his government to start measuring happiness.[47] And at his second inauguration in 2013, President Barack Obama reprised Jefferson: "That is our generation's task–to make . . . life, and liberty, and the pursuit of happiness real for every American."[48]

The Ant and the Grasshopper

If you had to suppress a giggle at the mention of Gross National Happiness–or perhaps you didn't even bother to suppress it–you're not alone. Happiness seems like cotton candy, pink and fluffy. It calls to mind a laughing young satyr prancing about in some meadow while others hunker down to the serious business of life. Scholars try to make happiness more respectable by calling it "subjective well-being," but that doesn't make it any less fluffy.

What taints happiness? One problem is captured in Aesop's tale of the ant and the grasshopper. During the summer, the grasshopper sings and frolics while the ant toils to prepare for winter. Come winter, the grasshopper suffers in the cold while the ant sits comfortably fed in his lair. In the original Greek fable, when the grasshopper knocks on the door, the ant tells the grasshopper to go dance and shuts the door in his face.[49] In modern versions of the story–cleansed of unhappy endings–the ant takes the grasshopper in, and the grateful grasshopper realizes the error of his ways.

Most people would agree that it's the ant who is happier in the long run, even though it's the grasshopper who seems explicitly focused on happiness. In other words, short-term pleasure often leads to long-term dissatisfaction. That intuition underlies the psychologist's distinction between *hedonia* and *eudaimonia*. Pleasure-seeking hedonism is questionable, but maybe long-term eudaimonic life satisfaction is good.

But is that enough? Modern psychological tests of happiness, which policymakers increasingly rely on, ask questions about current mood and life satisfaction so far. According to these measures, happiness depends on the present and the past.[50]

Yet the future is where all of our potential happiness lies. If happiness is present mood and present satisfaction, then efforts to increase happiness will tend, grasshopper-like, to focus on today, not tomorrow. This is exactly what is recommended by the recent spate of happiness literature. Take *The How of Happiness*, in which leading positive psychologist Sonja Lyubomirsky lists 12 Happiness Activities and 5 Hows Behind Sustainable Happiness. The latter include positive emotion; variety in life; social support; motivation, effort, and commitment; and habit. The first two are clearly present-focused, and even the last three are upon closer reading. Consider motivation, effort, and commitment. These sound more ant than grasshopper, but Lyubomirsky gives the section only four pages of which just one is dedicated to the question "What if you're too busy?" Her answer is to do simple things that take no additional time, such as "observing your job, partner, and children with a new, more charitable and optimistic perspective, saying a kind word to your spouse, distracting yourself when you find yourself dwelling on something, uttering a short prayer before a meal, smiling at strangers during your commute, empathizing with someone who has hurt you, and so on."[51] Nothing in the section suggests that motivation, effort, and commitment require motivation, effort, and commitment. Like so much positive psychology, all Lyubomirsky recommends are fleeting attempts to improve one's present mood.[52] There is nothing about practicing habits of courage, diligence, or integrity that might enhance future happiness. And in the course of her 350-plus pages, only ten are devoted to creating more happiness for others through kindness. Even there, the main concern is how kindness promotes one's own happiness, not others'. So, according to one of the world's foremost happiness experts, happiness is not about laying the foundation today for happiness tomorrow or happiness for others, but to do more as pop musician Bobby McFerrin sang: "Don't worry, be happy."

Happiness may be the intended end of all human effort, but as Aristotle recognized long ago, it is a byproduct of other activity. You might be as happy as a grasshopper today, but it's only through careful preparation that happiness lasts through the winter. McFerrin is wrong. If "the landlord says your rent is late,"[53] the right response isn't not to worry. It's to move in with a relative, live on a tighter budget, get another job, upgrade your skills, apply for public assistance, or do anything else that would sow, antlike, a future in which there would be less cause to worry.[54]

Measure for Immeasure

At first glance, randomized controlled trials, social enterprise, and happiness as an objective have little in common. The first is a research methodology, the second a type of organization, and the last a policy goal. Yet these three concepts share traits that recall the problems of technologies and packaged interventions.

For one thing, they all make a fetish of measurement. Randomistas look down on knowledge not gained through quantified experiment. Social enterprises are pushed to reduce their impact to a number. Happiness gained currency only when social scientists developed metrics for it. And whether it's vaccines injected, laptops issued, loans disbursed, or ballots submitted, one reason why packaged interventions are popular is that they are easy to count.

Measurement undoubtedly helps us verify progress. There's a danger, though, of worshipping the measurable at the expense of other key qualities. We can know the number of mobile phone accounts, but we can't know how many life-changing conversations they've carried. We can count votes, but we can't tell how many citizens will risk hazards to protest injustice. Someday these intangibles might be quantifiable, but even then, much will remain unmeasured. As the saying goes, "Not everything that can be counted counts, and not everything that counts can be counted."[55]

If so, it's important that we acknowledge here and now that important but numberless qualities will always exist, and that we account explicitly for that fact in our decision-making. Unfortunately, in our world of big data, we are losing sight of bigger wisdom. As more kinds of numeric data become available, we focus only on the numbers and neglect qualities that don't come with measurable outcomes.

Technocrats like to say that "if it can't be measured, it can't be managed," but this is simply not true. Most of us manage our relationships with friends and family without measurement. (And you'd worry about anyone who needed metrics to manage relationships.) Many countries have experienced dramatic economic growth well before they have had a system of national accounts.[56] Surely, Homer thought of his *Iliad* as much more than 15,693 lines of dactylic hexameter. The important thing is to establish meaningful goals first, whether or not they can be measured. Where direct metrics don't exist, there might be indirect proxies. And where there aren't proxies, there should be a judicious weighing of measurable and unmeasurable factors. It's foolish to neglect metrics where they're available–but to think that only what's measurable is meaningful is pure sophistry.

The Tech Commandments

The problem with measurement obsession is the obsession, not the measurement. The drive for lower-cost books squeezes out all but best sellers. A mania for RCTs crowds out complementary approaches. Social enterprises distract from other paths to charitable action. Near-term happiness diverts us from long-term foundations. A tunnel vision on technology steals attention from nontechnological essentials.

In the hype surrounding these technocratic approaches, certain biases appear and reappear constantly. They are the distortions of our technological and technocratic age–what could be called the Tech Commandments:

- **Measurement over meaning:** Value only that which can be counted.

- **Quantity over quality:** Do only those things that affect millions of people.

- **Ultimate goals over root causes:** Focus narrowly on the end goal to ensure success.

- **Destinationism over path dependency:** Ignore history and context, and take a single hop to the destination.

- **External over internal:** Do not expect people to change; instead, focus exclusively on their external circumstances.

- **Innovation over tried-and-true:** Never do anything that has been done before, at least not without new branding.

- **Intelligence over wisdom:** Maximize cleverness and creativity, not mundane effort. Use intelligence and talent to justify arrogance, selfishness, immaturity, and rankism. (Rankism is abuse, humiliation, exploitation, or subjugation based on any kind of social rank.[57])

- **Value neutrality over value engagement:** Bypass values and ethics by pretending to value neutrality.

- **Individualism over collectivism:** Let competition lead to efficiency; avoid cooperation, which breeds complacency and corruption. Any inhibition of individual expression, including compromise to support the common good, is the same as oppression.

- **Freedom over responsibility:** Encourage more choices; discourage discernment in choosing. Any temperance of liberty, including encouragement of responsibility, is tantamount to tyranny.

This is an exaggeration, I'll admit, but not an extreme one. I've been in hundreds of discussions about global poverty with academics, entrepreneurs, nonprofit staff, program officers, and government ministers. With striking regularity, someone will invoke some version of these points to justify their pet intervention with smug certainty of its power. Technocratic zealots aren't satisfied by seeing their point of view acknowledged; they want it to prevail. In this, they call to mind Larry Ellison, the cofounder and CEO of Oracle, who once said that he modeled his business tactics on Genghis Khan. "It is not sufficient that I succeed," he said. "All others must fail."[58]

Belief in the Tech Commandments isn't limited to technologists, packaged interventionists, or devotees of RCTs, social enterprises, and happiness. Nor is it confined to any one group.[59] It permeates social-cause circles on the political left and right, in the private and public sectors, among secular philanthropists and religious charities.

The Tech Commandments aren't easy to counterbalance, because they are not wrong in and of themselves. It would be pointless, even dangerous, to argue against metrics, innovation, or freedom per se. But as the Delphic oracle advised, *medem agan*–nothing in excess. Balance is utmost.

Teaching to the Test

The imbalances of the Tech Commandments creep into our systems little by little. Over time, however, they can snowball into major crises.

One example of this tendency is teaching to the test. A struggling school system prompts a narrowing of goals. The focus shrinks to reading, writing, and mathematics, which seems sensible enough and incurs little partisan controversy. Then, to measure progress at large scale, standardized tests are used as benchmarks. Pretty soon, raising the metrics becomes the only goal. Under pressure to increase scores, schools turn to quick fixes: technologies and methodologies that drill students in minor variations of common test questions. However, rather than fostering curious, productive, well-informed, and well-adjusted citizens, the mindless drilling erodes students' motivation to learn.

Meanwhile, these changes prompt parents with means (as well as parents with vouchers) to send their children elsewhere–to private and charter schools. This response institutionalizes a two-tiered system that only aggravates the original problem. What started as an attempt to save a challenged school system is undermined by the technocratic over-emphases on measurement, large scale, external change, individual choice, and supposedly value-neutral change.

This pattern generalizes beyond education. A struggling public effort prompts a narrowing of goals. The focus shrinks to improving health care, education, or economic output, which seems sensible enough and incurs little controversy. Then, to measure progress at large scale, the goals are benchmarked by mortality figures, average years of schooling, or income and GDP. Pretty soon, raising the metrics becomes the only goal. Under pressure to perform on the metrics, governments, donors, and civil society turn to quick fixes: technologies and methodologies that supposedly raise benchmark scores. However, rather than fostering independent, productive, neighborly citizens, packaged interventions delivered from the outside erode communities' own capabilities.

Meanwhile, high-status communities disregard the inconvenient aspects of the same packaged interventions they peddle. This response institutionalizes a two-tiered system that only aggravates the original problem. What started as an attempt to cause positive social change is undermined by the technocratic over-emphases on measurement, large scale, external change, individualism, and supposedly value-neutral change.

The Dimming of Enlightenment

I want to be clear that I'm not attacking technocratic goals in and of themselves. Technocratic ideas have become popular because they have profoundly altered human civilization in some very positive ways.

If you plumb the history of technological invention and large-scale social change, a lot of it can be traced to seventeenth-and eighteenth-century Europe and the historical period known as the Enlightenment or the Age of Reason. The Enlightenment saw an explosion of intellectual activity and laid the basis for the Industrial Revolution. Everything from steam engines to seaworthy clocks, from telescopes to barometers, emerged from that time. But the Enlightenment was much more than a burst of technology. Books about the era would fill entire libraries, but to compress it all into a Facebook post: In scholarship, the Enlightenment ushered in the reign of science and reason over superstition and dogma. In culture, it exalted meritocracy and pluralism. In economics, it gave backing to property rights and national growth. And in government, it brought down dictatorships and gave birth to democracy.

These ideas were reactions to the prevailing unwisdoms of autocracy, imperialism, superstition, prejudice, and economic stagnation. Enlightenment ideas served as a counter to dogmatic rituals and monarchies that thrived on the backs of an uneducated, untitled population. Scholars of the Enlightenment routinely say that the period gave birth to the idea of *external progress* as something that was both desirable and systematically achievable for humanity as a whole.[60] That is in contrast to the world before the Enlightenment both in and out of Europe–a time when it wouldn't be too gross a generalization to say that human brainpower tended to focus on internal change. Ancient Greek philosophy and Judeo-Christian tenets ruminated on personal virtues. Confucian principles emphasized social harmony and respect for hierarchy. Indian religions stressed karmic forces and spiritual advancement.

Before the Enlightenment, major civilizations came and went on all the continents except Antarctica. None of them, though, built anything like the rich intellectual edifice on which the modern world was constructed.[61] History has its twists and turns, but you can draw a straight line from the ideas of the Enlightenment to the contemporary world. Isaac Newton and others paved a highway for science and technology with their explorations into the laws of motion and electromagnetism. Baruch Spinoza and Jean-Jacques Rousseau laid the philosophical cornerstones of modern democracy. John Locke's arguments for property rights and Adam Smith's analysis of markets undergird contemporary capitalism.

And thanks to those foundations, some portion of the world gained more prosperity, justice, dignity, freedom, happiness, and peace than any civilization that came before. In 2006, global GDP was $50 trillion, with the rich OECD countries producing three-quarters of that dizzying sum. Most developed-world citizens had their basic needs met. Few struggled for food or shelter. Life expectancy at birth was seventy-seven years (well above the fifty-five-year life expectancy for the least developed countries, or the thirty-nine years of, say, Massachusetts residents in 1850). Most OECD countries were democracies with rule of law and basic human rights protections. While some were engaged in war abroad, peace reigned at home. And the Global Values Survey found that citizens of the richest countries were consistently happier than their counterparts in poorer ones.[62] So 2006 was a great year for beneficiaries of the Enlightenment. It seems only natural that we should keep doing more of a good thing.

Yet, just a few years later, things look much darker. We remain mired in an economic slump that is dragging down both spirits and bottom lines. In 2013 the world burned 33 billion barrels of oil (about half of them in the rich countries), chugged 207 billion liters of soft drinks, and cut down 1.4 million acres of Amazonian rain forest, and all of these things contributed on a colossal scale to pollution, climate change, a looming resource crunch, and poor health around the world.[63]

What's more, the world's richest, freest, happiest people are the ones who are most responsible for these problems. On a per capita basis, Americans consume as much as thirty-five times the natural resources of their developing-country peers.[64] And the world's financial troubles can be linked to the excesses of Wall Street, where people live like demigods and buy their own justice. It seems that prosperity, justice, dignity, freedom, happiness, and peace do not guarantee themselves–either forever or for everyone. If anything, it's our success with these desirables that increasingly infringes on our neighbors and on our own future.

How could something so good have gone so wrong?

Essentially, we took the progressive ideas of the Enlightenment and forged them into a rigid technocratic orthodoxy. We are unable to entertain alternatives to tech-driven, capitalist, liberal democracy, so we pronounce it the ultimate salvation.

But if Adam Smith's invisible hand spurs economic growth, it also pickpockets from the commons. Moral relativism permits plural virtues but also plural vices. Meritocracy rewards talent and diligence but neglects the collective responsibility to nurture those traits in everyone. We rationalize our faith on the basis that it leads to the common good, all the while winking at the foibles it indulges.

Minor character flaws then weave a tragic destiny. To be sure, a value-free pluralism is better than monarchic oppression, but it is still not enough to ensure the public good. Wall Street bankers and global mining companies are un-oppressed, and free to trade pretty much as they wish, but in the daily choices they make between protecting the public good and fattening their bank accounts, does anyone truly believe they are choosing for the good? And if not, why do our dearly held convictions still support them? Why do we get excited about innovations such as high-speed electronic trading, even though they achieve little apart from amplifying greed and reinforcing undeserved advantage?

When Sarkozy announced his quality-of-life commission in 2009, he said, "We're living in one of those epochs where certitudes have vanished. . . . [W]e have to reinvent, to reconstruct everything. The central issue is [to pick] the way of development, the model of society, the civilization we want to live in."[65] Whatever his flaws, Sarkozy put his finger on the core quandary of modern global society. We need a better story about what progress is and where we go from here. The Enlightenment served its purpose, and many of its technocratic values are worthwhile. But now we're clinging.

Unfortunately, Sarkozy's commission was a twenty-five-member group composed entirely of economists. To be sure, they were eminent economists, led by Nobelists Joseph Stiglitz and Amartya Sen.[66] But, while economics in the abstract can be construed as a science of human well-being, economics as it is practiced today is a monstrous hydra of rational-agent models, linear utility functions, oversimplified regressions, dollar-based metrics, conflation of meaning with measurability, and can't-help-itself support of free markets–and its many heads reassert themselves despite routine failures.

Again, it's not that technology, packaged interventions, RCTs, social enterprises, happiness, scalability, measurability, and technocratic ideas in general are bad in and of themselves. Rather, the trouble is cultism and imbalance. New vaccines are good, but not while health-care systems go unfunded. Educational technology might be helpful, but not if good teachers and institutional support are lacking. Elections are great, but not if social norms and government institutions don't support democracy. Technocratic means might be a part of the solution, but with so much attention on them, who's working on the other parts?

Balance is utmost, but balance is difficult even to talk about in a world of polarized sound bites.[67] Novel, measurable, large-scale, turbo-charged, value-free, market-oriented packaged interventions for freedom-drunk, goal-driven, meritocratic individualists dominate our notions of social change. This creed has been terrific so far for those of us who have benefited. But the world's persistent challenges and imminent crises suggest that what got us here won't take us further. For a more enduring humanity, we need a better narrative of progress.

PART 2

Amplifying People

CHAPTER 6

Amplifying People

The Importance of Heart, Mind, and Will

I first met Rikin Gandhi in 2006. He was a software engineer at the time, and Digital Green was still in our future.

Gandhi was a dreamer whose chiseled facial features belied a methodical intensity. He had always wanted to be an astronaut, so he studied the biographies of men and women who'd gone to space. He could recount the intricate details of every Apollo mission. And he knew that one of the best ways to be selected by NASA was to have an engineering degree and to be an Air Force pilot. When we met, he had just completed a master's in aeronautical engineering at MIT and was working at Oracle while waiting to be admitted to the US Air Force Officer Training School.

But the wait was long, and while he waited, he also noticed a theme in those space-traveler bios. "Astronauts get a chance to see our world as the small blue marble that it is," he told me, "and they come back with new love for humankind and for the earth itself." A friend of Neil Armstrong's once explained, "You understand that you're a short-term phenomenon, like the mosquitoes that come in the spring and the fall. You get a perspective on yourself. You're getting back to the fundamentals of the planet."[1] Some, such as Armstrong himself, retired from NASA and went into farming. Others, such as John Glenn, took up public service. And all of this awakened something within Gandhi.

Gandhi decided to merge the astronauts' love of land and people by working on behalf of poor farmers in India, where his parents came from. And, in the fashion of so many engineers who want to support social causes, his plan was to apply technology. He wanted to run "rural telecenters"–something like Internet cafés but meant for poor communities. Proponents envisioned that the world's villages and slums, once connected, would have access to better health care through telemedicine, better education through distance learning, and better agriculture through online research.

In India, telecenters were held out as a panacea. Through them, entrepreneurs, academics, and policymakers thought they could spread the success of the technology sector to the vast rural population. M. S. Swaminathan, hailed as the father of India's Green Revolution, wanted to put "Village Knowledge Centers" in each of the country's 600,000 villages.[2] The government also initiated its own project to set up "Common Service Centres" nationwide.[3] Professor Ashok Jhunjhunwala, a member of the prime minister's science advisory committee, claimed that rural telecenters could double household incomes in remote villages.[4] Still others argued that there ought to be universal access to the Internet, and that it ought to be considered a human right.[5]

Gandhi was caught up in the excitement but shrewd enough to sense that improving lives through digital services would not be easy. He wanted to speak with people who had direct experience. He found me through a report on telecenters I had written with a colleague, Rajesh Veeraraghavan, and the three of us arranged to meet.[6]

Unfortunately, the research Veeraraghavan and I had done showed that telecenters rarely met their business or social-impact goals. I had visited about fifty telecenters throughout South Asia and Africa, and the vast majority of them saw little footfall. Most telecenter operators didn't have the marketing skills to sell their services, and their would-be customers saw little value in impersonal medical advice, teacher-less learning, or academic papers about agronomy.

In the face of these problems, supporters proposed further technological fixes. If telecenter operators lacked skills, they would design online communities to share best practices. If rural patients wanted to see real physicians, video teleconferencing was the answer. If there were barriers of language and literacy, they called for more user-friendly content tailored for local needs, translated into local languages and shot as videos that didn't require reading. And on and on.

But little of that addressed the underlying problems. The very people whom the telecenter advocates wanted to reach–people with low incomes and little education–were the ones least likely to be able to pull themselves up by the bootstraps of abstracted knowledge or anonymous communication. Even well-educated self-starters prefer the formal structure of classes, insist on face-to-face appointments with doctors, and seek professional advice from warm-blooded mentors. All of this was lacking from the telecenter experience.[7]

The few telecenters that weren't total failures fell into three categories. Some rebranded as Internet cafés, gave up on social causes, and succeeded as commercial businesses.[8] Some saw a demand for computer-literacy certification and turned themselves into IT training schools that mostly catered to the relatively well off. Some were indefinitely supported by dedicated nonprofit organizations that viewed telecenters as a part of their charitable work because they buttressed existing programs. Telecenters amplified the underlying intent and capacity of their operators, but in themselves they did little to address deep social challenges.

As Veeraraghavan and I relayed these lessons, Gandhi's shoulders sagged. He had invested a lot of hope in his project. He was crestfallen.

But not destroyed. Gandhi kept in touch, and each month for several months, Veeraraghavan updated me about their conversations. Much of what they discussed came down to this: If widespread dissemination of cookie-cutter interventions such as telecenters weren't the right way to address social causes, what was?

The Birth of Digital Green

After a few months, Veeraraghavan came to me and said, "I think we should hire Gandhi. He's committed to supporting smallholder farmers, he has great technical skills, and he's tenacious. If you agree, I have an idea he could try out." We brought Gandhi on board to find a way to use video to help teach farmers. He came up with Digital Green, which went on to become one of the lab's most effective projects.[9] Digital Green, in turn, provided ample support for technological amplification and influenced my thinking about packaged interventions.

When Gandhi began, we put him in touch with a small nonprofit called Green Foundation. It performed agricultural extension–training for farmers–in a block of villages about two hours south of Bangalore. For the next six months, I barely saw Gandhi. Most weeks, he stayed in the villages. He worked alongside Green Foundation staff and got to know farmers in the area. Every once in a while, he'd pop into the office, and we'd talk about what he had been up to and what he could do next. One week, he told me, "I think I now understand Green Foundation's agriculture programs, so I'm moving on to capturing their advice on video." Another week, he told me how he'd tried everything from "how-to videos featuring staff, how-to videos featuring local farmers, and testimonials by respected villagers. I'm also trying to see if local entertainment like children singing folk songs will draw people in." Video in hand, he set up events at people's homes, viewing sessions in the local school building, and screenings by laptop in the very fields where people were farming. "I've tried everything," he told me. "Here's a photo of a television I set up in the middle of the main road. I wanted to see who would show up."

Eventually, he settled on a set of practices that seemed to work. The core concept was to use how-to videos featuring local farmers as teaching aids during scheduled weekly screenings. At the session, a village resident–sometimes accompanied by Green Foundation's agricultural extension staff–would hold a discussion. "It's important that the videos feature local farmers so that the audience can immediately identify with them," Gandhi explained. "What we're doing is very different from the farming shows on TV," which the farmers watched but mostly disregarded. In contrast, Gandhi's actors spoke the same dialect, wore similar clothes, and lived in the same environment as the viewers. "Also, an active mediator is critical," he told me. "The farmers engage more when provoked into discussion, and if there's someone there to take questions." And so, Digital Green was born.

At that point, we ran a controlled experiment to verify the initiative's impact. The trial ran for a year and a half. Eight villages ran Digital Green, another eight ran a classic style of person-to-person extension known as "training and visit," and four villages ran Poster Green–just like Digital Green, but with lessons in poster form.[10] Gandhi and Green Foundation staff painstakingly tabulated what happened at every video session and went field to field in the villages to record adoption of practices. The results showed that Poster Green performed quite a bit better than classic extension programs, but that the advantage faded after about five months. The audience grew bored of posters, yet the farmers often needed the content repeated to remember it. Digital Green, however, outperformed Poster Green even at its best, and its performance never flagged. Farmers were willing to watch what amounted to the same content as long as we rotated in new videos with different farmers in them. Overall, Digital Green caused seven times more adoptions than classic extension, and it was ten times more cost-effective. With Digital Green, a single extension officer–whose salary was the expensive part–could serve more villages than he could have meaningfully addressed one-on-one. With video as a focus, it was possible to hold discussions with many farmers at once.

In Digital Green, we see the Law of Amplification's positive power. Offering videos to poor, nonliterate farmers is largely meaningless by itself. That's why little of the agricultural content offered through telecenters ever sees much use, and why broadcast programs on Indian public television have scant impact. But farmers are swayed by in-person interactions with peers and extension officers. Digital Green's videos amplify those human-to-human interactions by making them more memorable and enabling more of them at once. They allow a partial substitution of experts with nonexpert facilitators. They leave a stronger impression on farmers.[11]

The Three Habits of Highly Effective Technology Use

Digital Green shows us that the best use of packaged interventions is selective and targeted. Its lessons can be condensed to three rules:

Rule 1–Identify or build human forces that are aligned with your goals. Even without digital technology, Green Foundation was committed to farmers and capable of supporting them. For packaged interventions to have positive impact, they need a positive human force to amplify.

Rule 2–Use packaged interventions to amplify the right human forces. Gandhi observed what Green Foundation was already doing and used technology to amplify its work. It's also possible to amplify the impact of unorganized social trends. In Kenya, for example, a mobile money transfer system called M-PESA famously increased the flow of money from urban to rural areas because there was already an underlying culture of urban migrants sending cash back home.[12]

Rule 3–Avoid indiscriminate dissemination of packaged interventions. Digital Green doesn't work without a strong partner that has rapport with farmers. And Digital Green didn't branch out into, say, children's education, because its partner organization had no expertise in that area. Seeking mass dissemination of technology for its own sake is a waste of resources and often counterproductive.

Tech-centric social projects most often violate the last rule. It's tempting to think of Digital Green as an all-purpose tool for knowledge dissemination. Some donors and partners see this potential in Digital Green just as they previously did with telecenters and currently do with mobile phone platforms. It's an understandable impulse–why not use the full potential of the technology to address health care, home economics, governance, nonfarming vocational training, and everything else in a single stroke?

But that's technological utopianism. There's a wrong way and a right way to work with the underlying intention.

The wrong way is to believe that the packaged intervention itself solves the problem. Say we collaborate with Partner X whose sole expertise is in agriculture. Partner X, however, notices a new demand in the communities it works with: Expecting mothers want advice during pregnancy. This appears easy to solve with videos, so Partner X looks up maternal health information online, produces new videos, and screens them for the community. But the technology doesn't substitute for Partner X's lack of expertise in medicine. If, after watching the videos, the mothers have questions, Partner X can't answer them. The staff has no real knowledge, and websites contain conflicting advice. If a woman has a difficult labor, Partner X doesn't have the connections to know which local clinic is best suited to help her.[13] There's a heightened chance that Partner X unintentionally provides bad information, which could end in tragedy. Even assuming Partner X is comfortable with what is starting to look like an ethical breach, over time its lack of expertise is noticed. The local community starts losing faith not only in the health-care information, but also in Partner X and in the original packaged intervention.[14]

To counteract these problems, Partner X could hire medical staff, train its staff in health care, or set up collaborations with local health-care organizations. Any of those options would be the right way to go. But these are deep changes to the kind of organization Partner X is, not just to the intervention itself. In fact, today Digital Green is slowly expanding into health care, but not by asking its agricultural partners to take on health videos. Instead, it works with new partners who are experts in health care. Digital Green's video sessions are effective exactly in those topics where its institutional partners have expertise.

Scaling Up

In 2008 Gandhi spun off Digital Green into its own nonprofit organization, and Veeraraghavan and I joined its board. From the outset, our model was to amplify existing positive forces in agriculture.

What does this mean at larger scale? Though Gandhi has absorbed a lot of agricultural knowledge over the years, he is by no means an expert. Nor are most of us on the board or among Digital Green's senior leadership. Our strengths are in nonprofit management, technology, and international development. None of us has special relationships with large numbers of smallholder farmers.

So Digital Green seeks out partner organizations that are deeply, capably invested in aiding poor rural farmers. Videos amplify their mission and expertise.

Over the years Digital Green has worked with a range of nonprofits and government entities. It has extended its reach to 5,000 villages through thirty partner organizations. It works in nine Indian states as well as parts of Ethiopia and Ghana. Over 400,000 people have viewed at least a few of the 3,000 videos produced in one of twenty languages. Some have seen their income more than double.[15] And on the whole, we find just what amplification would predict: The stronger the partner, the better use they make of Digital Green, and the greater the impact on farmers. In other words, Digital Green's success at large scales is entirely dependent on the presence and reach of good partners.

The flip side of this is that there is no Digital Green without a partner. Many tech-centric projects hold themselves out as comprehensive solutions, or, if they're slightly more careful, as a primary solution that requires a little human support. In contrast, Digital Green is keenly aware that its staff and its partners do the hard work of building relationships with farmers and identifying appropriate agricultural practices. Like all well-implemented packaged interventions, Digital Green piggybacks on human forces, which are the primary agents of change.[16]

Because Digital Green doesn't substitute for agricultural organizations where they're either missing or dysfunctional, Gandhi doesn't lobby to make video access a universal right for farmers. Nor is he in a rush to provide Digital Green content to every farmer on the planet via mobile phones or drone-delivered iPads. Not while so many countries lack institutions that farmers trust.

Digital Green's method of working captures the right way to apply technologies: It identifies or builds organizations and social trends aligned with its goals and then targets packaged interventions to amplify their impact.

Heart, Mind, and Will

Since partners are essential, they must be chosen wisely. In my experience across a range of projects beyond Digital Green, I've found that three qualities make a good partner: good intention, discernment, and self-control, or what I've come to think of as *heart, mind, and will.* These qualities are what technologies amplify, so when those human qualities are at their brightest, packaged interventions also shine.

With Digital Green, heart, mind, and will start with Gandhi. He wants to help farmers (intention); he has solid technical skills and seeks constructive feedback (discernment); and he perseveres when his interests are on the line (self-control). His devotion to developing-world farmers is unwavering; he's a shrewd CEO for Digital Green; and his employees marvel at his nonstop work ethic.

Heart, mind, and will are also what Digital Green seeks in its implementing partners: commitment to supporting smallholder farmers (intention), agricultural expertise as well as knowhow for building rapport with farmers (discernment), and the tenacity required to follow through (self-control). "We work hard to vet our partners," Gandhi says, "to make sure that they're good at what they do, because Digital Green's success depends on them."

Beneficiaries, too, need heart, mind, and will. Farmers must have some desire to improve their own lives (intention), basic agricultural knowledge and the ability to pick up new practices (discernment), and the willingness to expend effort to learn (self-control). Where they need support, Digital Green's partners provide it.

Two things about heart, mind, and will are worth pointing out. The first is that, at the very least, they are necessary *complements* to packaged interventions. Even vaccines and medications–which are as close to a complete solution as packaged interventions ever get–require the heart, mind, and will of willing patients, caring nurses, and expert doctors.

The second point is that heart, mind, and will are also the *cause* of packaged interventions, both in their invention and provision. The Digital Green methodology would never have been invented without Gandhi's dedication, discernment, and self-discipline. And Digital Green would not be implemented properly without those same traits in partners. The same can be said for vaccines. Their inventors require heart, mind, and will. And so do the governments, large foundations, and multilaterals like the World Health Organization that disseminate them.

It may seem obvious that packaged interventions work best when leaders, implementers, and beneficiaries have great amounts of heart, mind, and will. But plenty of smart, influential people behave otherwise, as if spreading technologies and packaged interventions indiscriminately were the way to cause social progress. To do so, though, is to make an idol of the easy part and neglect the rest–the finding or nurturing of the right heart, mind, and will.

Amplification Under Conflict

Sometimes leaders, implementers, and beneficiaries aren't aligned. What then?

Soon after recruiting Gandhi for Digital Green, Veeraraghavan left for his own quest: a PhD at UC Berkeley's School of Information. He returned to India for his fieldwork, where he studied how digital technology might support democratic governance, focusing on the National Rural Employment Guarantee Act (NREGA). "NREGA is a national program in which people below the poverty line are paid a fixed wage," Veeraraghavan explained. For about US$2 a day, up to a hundred days a year, workers do "menial jobs like road building that are chosen by local village governments. The idea is to guarantee some amount of work for India's poorest people while also building local infrastructure." But as with any national program in the world's messiest democracy, what happens on the ground frequently differs from what policymakers intend. Village-and block-level leaders often submit falsified work records and pocket the payout. The intended beneficiaries don't get paid, and infrastructure doesn't get built.

In the southern state of Andhra Pradesh, however, government leaders were determined to make NREGA work. "They knew about corruption within their own administrations," Veeraraghavan says, "so they did two things to address it. First, they imported a concept called 'social audit' that was a part of NREGA legislation." Social audit–championed by Mazdoor Kisan Shakti Sangathan (MKSS), a nonprofit that has long worked for the civil rights of low-income workers–is, Veeraraghavan says, "a process by which local citizens publicly review government books to make sure that government projects are being implemented without corruption."[17] In MKSS's home state of Rajasthan, they implement intensive processes. Village-wide meetings, door-to-door verification, and follow up with government all ensure that improprieties are corrected. But MKSS routinely deals with recalcitrant local officials who resist opening up their books. And many a balance sheet is filled with illegible scribbles. According to Veeraraghavan, in Andhra Pradesh "the government also installed an online system in which all NREGA activity was supposed to be logged. The data was open to public scrutiny, and corruption was supposed to decrease."

Veeraraghavan found, however, that the effect of digitization was mixed. In some cases "social audit and digital record-keeping shined a spotlight on corruption, and corruption came down drastically." But in other cases, corruption and inefficiency continued and were even exacerbated by the technology. For example, a plan to monitor low-level officials by GPS foundered because upper management was unwilling to enforce penalties for noncompliance. One manager said, "If everybody is misbehaving . . . how many [officials] can I suspend?"[18] Low-level bureaucrats would occasionally shift the blame for their corruption, telling villagers that "the computer has not assigned them work." And sometimes, Veeraraghavan says, villagers not only failed to hold bureaucrats to account but colluded with them to cheat the system: No one did any work, but everyone agreed to log false records so as to be paid.

Amplification is the best explanation yet again, despite the conflicting politics. Veeraraghavan says that for the government leaders, the technology was "an amplifier of their intention to curb corruption." But the technology also amplified the intention of local officials and villagers to game the system. Overall, much of the leakage was sealed, but full accountability wasn't achieved.

The challenges of democratic governance require a far more patient approach of a kind discussed in Chapter 9. For now, it's worth going back to education, where we have more immediate power to apply technology well.

Raising Digital Natives

No sooner had I turned away than I began to hear the music again. It was the synthesized soundtrack that accompanied Exit Path 2, a video game I had grown all too familiar with over the previous weeks. Vincent, a bright, energetic nine-year-old, who wore a baseball cap that would point every way but forward, had reactivated the game on his laptop.[19] Minutes ago I had scolded him for starting the game in the first place. He was supposed to be using a piece of educational software called Scratch that taught children the basics of programming.

Vincent was a student at the Technology Access Foundation (TAF) in Seattle, where I taught an afterschool course during the spring of 2012, following my return from India. I was curious about what I could do to support social causes, if simply inventing and disseminating new technologies wasn't the answer. At TAF, I hoped to verify a hunch I had: that teaching about technology was fundamentally different from providing new technology. That turned out to be true, but the real lesson was something else entirely.

TAF's founder and CEO is Trish Millines Dziko, one of the few black women in the 1970s to graduate with a computer science degree. Upon joining Microsoft in the mid-1980s, she saw firsthand that some racial minorities were dramatically underrepresented in the high-tech industry. Thus, TAF's mission is to "equip students of color for success in college and life through the power of a STEM [science, technology, engineering, and math] education." On Mondays and Wednesdays, I had two classes each with twelve third-to fifth-graders. Every student was allocated a laptop, and the curriculum called for hands-on activities to explore computer programming, audio editing, and robotics–topics that TAF's students weren't exposed to in their underfunded public schools. TAF's staff and I worked hard to come up with fun, engaging activities. Students produced their own YouTube videos, programmed interactive animations, and built Lego robots. But as much as the students enjoyed these activities, far greater pull was exerted by video games.

Children have a flair for sniffing out games and playing them behind the backs of supervising adults. They also have a sixth sense for educational content, and they avoid it like they avoid broccoli. In my class, the students preferred two-dimensional games in which colorful characters trot and leap through a cartoon-strip world. As innocent as these games were, even ardent video-game proponents–such as Jane McGonigal, author of *Reality Is Broken: Why Games Make Us Better and How They Can Change the World*–would be hard-pressed to identify their nutritional value. At best, they had some minor incidental benefit to hand-eye coordination.

Whatever the case, even the most beguiling educational software couldn't compete with the allure of Exit Path 2.[20] So there I was, hoping to increase children's understanding of technology, and technology itself was the obstacle.

My predicament at TAF was a version of the same problem that confronts parents and school systems around the world: How do we best prepare our children for an increasingly technology-rich world, while avoiding the perils of the technology itself? Once again, the first thing is to align human forces.

People First

My boss at TAF was Toyia Taylor, the program's manager. Taylor was at least a decade younger than me, but she commanded respect from children in a way I didn't. Spines straightened when she walked into the classroom.

Taylor's desk was just a few feet from my room, and after my second day at TAF, she pulled me aside. She hinted that I may want to be a little stricter with the students. She could hear commotion through the wall. This was the first time I had taught eight-to eleven-year olds, and I had quixotic notions about letting the students discover what they wanted to learn. What they discovered instead was that they could run all over me.

Taylor had concrete suggestions. A call-and-response clapping protocol to recover student attention. Five-minute timeouts for disruptive students. A meeting with her if that didn't work.

When I instituted the rules, the students naturally resisted. They tested me, and in quick succession, I had to send several students to the formidable Ms. Taylor. I felt bad about it, but I realized I had to overcome my kneejerk over-empathy for the children. Luckily, fifth graders are at an age where even the more disruptive ones still defer to adult authority. I found that anyone sent to Taylor didn't want to go back. Within a week, students were better behaved, and Taylor stopped casting concerned glances my way. Maintaining an attentive classroom atmosphere was the most difficult part of what I did at TAF, but it was a necessary foundation for learning.

There's nothing special about the rules I imposed. Experienced teachers have their own versions. What's notable, however, is that in a class about computers, what was required first was a change in people: good behavior from students and the willingness to discipline on my part. If technology amplifies human forces, then a poor outcome often means that the right human forces aren't in place. Where people problems exist, even the best technology will flop.

Does Not Compute

This principle applies to education broadly. To wonder what ails American education is to open a Pandora's box. It could be poverty in early childhood or school districts funded by inadequate property taxes. Maybe it's poorly designed incentives for teachers or elite flight into the private school system. The truth likely lies in some combination of these factors and more, but the problem is definitely not a lack of computers. Even people who think that more gadgets would help don't argue that US educational decline was caused by a decline of technology.

In America, much of our collective handwringing about education comes from comparisons with other countries. In the 2012 Program for International Student Assessment (PISA), American students ranked twenty-seventh in math and seventeenth in reading.[21] But while the United States as a whole may be losing its competitive edge, stronger students aren't sliding. At the annual International Math Olympiads, for example, where countries send their six best precollege mathematicians to solve problems that make SAT questions seem like 1+1, the United States regularly places in the top three.[22]

But as data from PISA show, high-scoring countries emphasize high-quality education for everyone, not just the elite. America, unfortunately, does poorly here when compared against thirty-three of the world's wealthiest countries. We have the third-lowest school enrollment rate (only 82 percent) for fifteen-year-olds, and we're ninth worst in educational disparity–scores vary particularly widely between well-off students and low-income ones.[23] We all know that our schools are unequal. Less acknowledged is that this inequality is responsible for our lack of competitiveness internationally.

If educational inequality is the main issue, then no amount of digital technology will turn things around. Secretary of Education Arne Duncan echoed the wishful thinking of many when he said, "Technology can level the playing field instead of tilting it against low-income, minority and rural students–who may not have laptops and iPhones at home."[24] This notion is misleading and misguided. Technology amplifies preexisting differences in wealth and achievement. Children with greater vocabularies get more out of Wikipedia. Students with behavioral challenges are more distracted by video games. Rich parents will pay for tutors so that their children can learn to program the devices that others merely learn to use. Technology at school may level the playing field of access, but a level field does nothing to improve the skill of the players, which is the whole point of education. Technology by itself only increases the gap between the haves and have-nots.[25] Educational technology scholar Mark Warschauer confirms that "the introduction of information and communication technologies in . . . schools serves to amplify existing forms of inequality."[26]

What the US education system needs above all isn't more technology, but a deliberate allocation of high-quality adult supervision focused on those who need it most. The specifics are daunting and complex, but this isn't a problem that technology can fix.[27]

The Proper Use of Educational Technology

Information technology's very power means we have to be careful with it. When I observed the other teachers at TAF, I found that while each of them had a unique set of rules, the effect in their classrooms was constructive learning. At first, this puzzled me. When their students opened up their laptops, they got to work; when my students opened up their laptops, they watched YouTube videos.

My problems began before the class started. When students arrived, they would pull out laptops from the cabinet and take them to their desks. For those who were early, I let them do as they wanted. They were there, after all, to become familiar with computers. It seemed sensible to maximize their time with the technology. Don't people learn best by doing?

But the result was that I would have to spend ten minutes at the start of class calling everyone to order. If I let them have dessert before dinner, they lost their appetite for the main meal.

Most of the other teachers didn't allow this. In fact, they carefully managed computer time. After consulting with other TAF teachers, I picked some rules that made sense to me:

- At certain times, such as when I was doing a demonstration, laptops must be closed.
- Laptops cannot be used before class, even if students arrive early.
- Laptop time in class must be used for class activities.

- Students who break any of the above rules twice in one day must go see Ms. Taylor.

The point was to constrain usage to educational ends. Sure, this was a computer class, but the goal wasn't to maximize screen time, it was to maximize learning. There are numerous theories about how children learn best, but it's clear that watching music videos isn't the best way to develop programming skills.

Veteran teachers will tell you that classroom rules should be introduced at the beginning of the school year in order to establish a productive class culture. I wasn't a veteran teacher, so I had to claw back privileges the students had gotten used to. It was a struggle. But once the students grew accustomed to my laptop rules, they (mostly) stopped trying to sneak in games and started focusing. I was happy to see that as students were freed from their inner yoke to games, they started asking questions about Scratch. "Can I add sounds to this?" "How do I make the balloon pop?" "What will make this character spin around five times?"

TAF taught me to be conscious and purposeful about laptop use. It was important to use technology strategically and to leave it out when it wasn't contributing to learning. Even Steve Jobs once admitted, "We limit how much technology our kids use at home."[28] Teachers, who must handle twenty, thirty, or forty children at a time, need to do the same. The advice is even applicable for older students. University professors, including me, increasingly prohibit device use in the classroom. "I'm a pretty unlikely candidate for internet censor," wrote Clay Shirky. "But I have just asked the students in my fall seminar to refrain from using laptops, tablets, and phones in class."[29]

Common Sense in Schools

In 2013, a friend asked me to help the board of the Northwest School, a private school in Seattle, think through their technology strategy. The school was known for its emphasis on the arts, its international student body, and its dedication to community service, but the board chair told me that they had been conservative about digital equipment. At one of their board meetings, I presented some of what I've covered in this book and then moved on to their specific questions.

The discussion was open, lively, and intelligent. Some of the board members were excited about computing and feared their children would be left behind. Others were anxious about potential distractions and liked the school's humanities bent. These are exactly the feelings that underlie so much of the debate about technology in our society. And they're valid. Emotions, though, are too blunt an instrument for carving strategy. They inevitably pose the question in terms of more or less, yes or no: Should every classroom have a smart board or not? Should we have WiFi or not? Should students all have laptops or not?

The real questions, though, require more precision, and the best way to think of them is to ask, What positive forces should be amplified? (And what negative ones, not?) I nudged the discussion toward specific educational goals and how technology might help achieve them. The school had a vibrant theater program, for example, which would benefit from good video production tools. It also came out that the school encouraged unique teaching styles, and that some teachers had wanted smart boards. The consensus was to install them when requested, but to recognize that not every classroom needed them. Some faculty liked to film their lectures and make them publicly available. That shouldn't be discouraged, everyone agreed, but the school's focus would remain on the students under its roof. A major effort to put lessons online seemed unwarranted. And then there was the inevitable question of campus-wide wireless access. As we discussed the educational goals it might serve, the need seemed to vanish. Most students carried smartphones and had PCs at home. The library had several rows of terminals with Internet access. None of the students were complaining that they couldn't do research online.

The biggest concern was whether to offer a computer programming class, which the school had never done. One father argued passionately that every profession increasingly benefited not only from the use of computers but also knowledge of how to program them. Some agreed. Another person hinted that the school's mission was broad and explicitly avoided vocational training. Others agreed. The debate went back and forth. Programming could be offered as an elective, but because faculty, student, and classroom schedules were jam-packed, something would have to be cut, and there was no agreement on what that would be.

At the end of the meeting, that question remained unresolved. I knew, though, that whatever decision they made, it would be right for them. It's the schools that work hard to maintain a strong learning culture, whose faculty and parents make important decisions together, and that put their educational goals first in making technology decisions–exactly the schools with strong heart, mind, and will–that technology's power optimally amplifies.

CHAPTER 7

A Different Kind of Upgrade

Human Development Before Technology Development

Think back to the Hole-in-the-Wall project. Children in poor neighborhoods were given free access to computers, and they figured out how to use them without any adult help. But they tended to do little other than play video games.

Those were kids, though. Maybe adults would be different. Such was the hope of my colleagues Sean Blagsvedt, Udai Pawar, and Aishwarya Ratan, who ran a project, inspired by the Hole-in-the-Wall, which they called Kelsa+. ("Kelsa" is the Kannada word for "work.") They wanted to see what adults who normally didn't work with computers would do if one was made available to them at no cost.

The team installed a PC in the basement of our office in Bangalore and connected it to the Internet. Then they held a meeting with our housekeeping staff, security guards, and technicians. The forty or so staff members were told that the PC was theirs to use as they wished as long as they abided by all laws and office policies. They were also told that the software would log any activity.

The PC got a lot of use, and after a few months the hard drive was stuffed with data. The researchers analyzed it, and what they saw combined the lessons of both Hole-in-the-Wall and telecenters. As with Hole-in-the-Wall, most of the staff–the men in particular–quickly picked up the basics of computer use. They learned from the technicians and security guards among them who had some familiarity with computers already. They browsed online, they sent each other brief emails, and they watched YouTube videos. One favorite activity was to replace the desktop's background image with portraits they took of themselves with the webcam. In surveys, the staff raved about the project. One of them said, "I felt so happy that day when we had the [meeting]. For the first time I touched a computer and did so many things without a mistake." His colleague mentioned, "Since the computer is here, we get awareness! Also because we can see the computer daily, my desire to learn and use it has increased very much." Another said, "In all my service, this is one of the best workplaces I have seen."[1]

Yet little of their computer use could be called productive. The most popular activities were watching movies and listening to music.[2] And their usage betrayed a shallow understanding of computer use. One funny habit was to attach ."com" to anything they searched for. When searching for the Indian movie *Pokiri*, for example, they would type "movie Pokiri.com" into the search engine, like a magical incantation. In any case, no one was learning skills that could help them in their home lives or careers.

Upgrading Technology Versus Upgrading Skills

Of the three researchers who started the basement PC project, Ratan stayed with it longest. Ratan is a woman of compact energy and a strong sense of social justice. Unlike many of the other researchers in the office, she knew the working-class staff by name, and though she was quite young, they all looked up to her.

Ratan wasn't satisfied with minor gains in computer literacy, so she convened another meeting with the staff and asked them what they wanted to learn from the computer. English was the most common response, so Ratan upgraded Kelsa+ with English instructional CDs.

Ratan administered English tests before the upgrade and three months afterward. The results were what you might expect. Although one person spent a lot of time with the software, most were casual users. Among the seven staff members who took both tests, there was an insignificant bump in average English proficiency.

But Ratan did have something else to be proud of. A security guard who was inspired by Kelsa+ enrolled himself in a data-entry course at a private training center. Every day after his shift, he used the Kelsa+ PC to practice what he had learned. After several weeks, he handed in his resignation. He had qualified for a job in data entry. He would take an initial cut in pay–as a seasoned security guard, he made slightly more than entry-level back-office workers typically did. But his growth potential was far greater. He might one day command a white-collar manager's salary that would dwarf anything he could have made in physical security.[3] He was happiest, though, about something less tangible: social status. "Today," he said, "I can stand up in front of my father and friends and say that I am no more a watchman, but I am doing a computer job."[4]

Kelsa+ shows that while giving away a packaged intervention does little on its own, a social aim can still be achieved if the intervention is paired with training. In international development circles, people often speak of "handholding" activities to encourage the best use of a packaged intervention, or "capacity building" to support organizational development. Microloans can be combined with financial education.[5] The sale of new seeds might occur within the context of proper extension. Vaccines are best delivered in the hands of well-trained health workers. Even the gift of a goat can be more useful with advice on its proper care.[6]

It's important to recognize, though, that a packaged intervention and the training for it are two very different things, just as owning a piano and knowing how to play it are. Training may require a physical object as a teaching aid, but it's much more than the provision of things. Teaching demands the effort and engagement of skilled people–things that can't be packaged.

Admittedly, training is expensive. But compared to a packaged intervention alone, training has far greater effect. On the one hand, about thirty-five people had access to the Kelsa+ PC in our office. They had the full range of "opportunities" that the Internet provides, but they got little of concrete value out of it. On the other hand, there was one person who sought training who experienced a dramatic upgrade in his earning capacity, social status, and life satisfaction.

Among my field-worn colleagues, you often hear that technology is only 10 percent of the solution. In fact, many projects that start out as technology giveaways evolve into instructional programs. Of the world's remaining telecenters, many survived by morphing into computer training schools. Even Sugata Mitra, the man behind Hole-in-the-Wall, admits that his open-access computers have more impact under the care of a teacher or school.[7]

The Amazing Students of Ashesi University

Kelsa+ reminded me of an experience I had before I went to India. In 2001 or so, my friend Nina Marini told me about an exciting venture she was about to help launch. She had recently graduated from UC Berkeley's School of Business, where she met Patrick Awuah, a Ghanaian American who was determined to start Ghana's first liberal arts college. Marini was inspired by his vision and had signed on as Ashesi University's founding vice president. I offered that maybe one day I'd teach there.

That day came earlier than I expected. In 2002, Ashesi took on its initial cohort of students and was in urgent need of someone to teach first-semester calculus. Marini asked me if I could fill in for one term, and I agreed.

When I arrived, there were two surprises. The first was that the class was wholly unprepared for calculus. A diagnostic test showed that two-thirds of the twenty-five students hadn't mastered basic algebra. Only a few of them could plot a straight-line equation on a graph. I wanted to deliver on what I'd promised, but in the one quarter I was allotted,

I couldn't squeeze in several years' worth of remedial math. The textbooks I had ordered were a poor fit, so I decided to develop a streamlined curriculum from scratch. I would shepherd the students from algebra through basic calculus, but focus narrowly on single-variable polynomials.

The second surprise was just how eager and motivated the students were. They were a teacher's dream come true. I had initially reserved a second track on trigonometry and exponential functions for a small group that was on firmer footing, and I announced on the first day that they would learn it through supplementary homework. But during office hours a long queue formed outside my door. One by one, the other students begged to be included in the second track. I gave in–it seemed wrong to decline so much aspiration.

The students worked hard. I assigned several hours of homework each night–meanwhile, they had other classes–but they were spurred on by the challenge. They absorbed the material at an amazing rate, as if they'd been starved all their life for the knowledge. By the end of the term, everyone understood derivatives and integrals and could perform calculus on polynomials. A good portion also mastered sines, cosines, and exponentiation as well. Final exam grades were all As and Bs, and one heroically achieved B–(about which more in a moment). I went home with the satisfaction that I had contributed in some small way to the students' education and professional growth.

I've stayed in touch with the students ever since. I returned to Ghana for their commencement in 2005 as well as for Ashesi's tenth anniversary in 2012. Today those students are programmers, entrepreneurs, and experts in a range of professions. Solomon Antwi is a consultant in international development. Andrew Tarawali is an investment adviser at an agricultural investment fund. Kweku Tandoh heads engineering at Rancard Solutions, one of Ghana's top software firms.

The students learned exactly what they'd signed up for: computer programming and business administration. But the experience meant so much more. Ashesi students aren't just capable users of technology.

They're leaders of technology and enterprise creation. If Kelsa+ shows that a little instruction is worthwhile, Ashesi demonstrates that deep investment in people is even more valuable.

Intrinsic Growth

In 2012, my former students and I had a reunion dinner at the fancy new Accra Mall in Ghana. As I reminisced with them and heard about their lives, it occurred to me that Ashesi graduates were members of a global elite. Despite their unique culture and history, most of them nevertheless shared the same opportunities with middle-and upper-middle-class people from around the world. And those opportunities were noticeably different from those of the average Indian telecenter customer, or some of the working-class parents of TAF students, or any of the dozens of other people I've spent time with in the rural and urban-slum developing world.

In my class, the math they learned was less important than the growth they experienced in heart, mind, and will. Linda shot me blank looks in class but forged her bewilderment into the knowledge she needed to ace the next day's quiz. Maame Dufie said she'd never thought she could learn calculus, but after doing well on the final exam, she knew she could do anything she put her mind to. When I polled some of my former students about what they learned most at Ashesi, no one in fact mentioned the specifics of calculating a derivative. Seyram Ahiabor mentioned that Ashesi's liberal arts education granted him versatility and "a lot of confidence in taking up challenges." Michelle Eghan learned that she had "the choice to question everything and not to accept things at face value." Isaac Tuggun said, "I knew it was going to be tough, but it was the best decision I ever made."

Tuggun was the student with the heroic B–I mentioned earlier: "I am from Nandom, a border town near Burkina Faso, in Upper West Region," he once announced to me, "one of the three deprived regions in Ghana." He was about ten years older than the other students, and from a comparatively less privileged background. "I lost both parents while still in secondary school," he said in his formal but fluent English. "My junior brother and junior sister were helpless and could not do anything to help themselves. There was no money. Even to get food was a problem. As a student I had to rear [a] few pigs and rabbits, a little poultry, selling cubes of sugar, cigarette, et cetera, all in small quantities." On top of these struggles, his legs had been permanently damaged in a childhood accident, and he had never been able to receive proper medical care for them. He walked on crutches.

After ensuring some independence for his siblings, he explained, he "spent a number of years seeking employment in Ouagadougou, the capital of Burkina Faso": "I did translation for business people and even government officials or agencies for a living. But that could not transform much of my life so I came back to Ghana in 2000. I was seeking both employment and an opportunity to study, and I got admission at Ashesi University." Tuggun was a bit of a legend at the school. Staff recalled the day he showed up at the admissions office with few formal qualifications. Turned away at first, he came back repeatedly to make his case. Eventually he was admitted on the basis of sheer charm, daring, and persistence. He even received a scholarship.

He was the lowest scorer on my diagnostic test and every subsequent exam, but he was also the hardest worker. He was often the only person in the library, plugging away at assignments after the other students had gone home. He frequently came to my office for extra help. He struggled to grasp each concept and handed in most of his assignments incomplete because he ran out of time. But he turned them in with a smile. And what he learned, he held onto. Tuggun earned that B–.

I kept in touch with Tuggun over email for a couple of years until, suddenly, I stopped hearing from him. He disappeared for seven or eight years. At Ashesi, they mourned the fact that he had never completed his degree.

Then, just as suddenly, he reappeared, right before the reunion dinner. Another classmate had run into him and put us back in touch. I met his fiancée at the dinner, and she helped arrange for Tuggun and me to meet in the northern city of Tamale, where nonprofits proliferate like casinos in Las Vegas. We both happened to have other meetings there. On the day of our appointment, I waited for him in the warm, dark air of an outdoor restaurant. A sturdy pickup truck–the kind favored by United Nations officials–drove up. When the passenger door opened, out came a pair of crutches followed by none other than Isaac Tuggun. He told the driver to wait for about an hour while we talked.

It turned out that Tuggun had worked his way up the ladder of an advocacy organization called the Ghana Federation of the Disabled. "I started as administrative secretary and rose to the position of administrator," he told me, "and then to national advocacy officer of the disability movement in Ghana." When I met him in Tamale, he was managing a joint project with DANIDA, a Danish aid organization.

Three Pillars of Wisdom

How did Tuggun go from itinerant interpreter to being chauffeured in the limousines of the foreign aid world? He displays several characteristics that many self-starters take for granted, but which are much rarer among involuntarily impoverished people. Some of these qualities come with a good education, which is an express advantage. But there's a tendency to think of education as being simply the ability to read, write, and do arithmetic, when an effective education involves so much more than academics. Conversely, many people lead satisfying lives because of their social skills, entrepreneurial drive, or force of personality without knowing a lot of science or history. In other words, even if education is one means to acquire the relevant traits, it's often less about the textbook content than something else that happens to come with formal learning.

What allowed Tuggun to succeed were the same heart, mind, and will that make packaged interventions work. Tuggun has good intention, discernment, and self-control–qualities similar to what Rikin Gandhi had in leading Digital Green, and which Digital Green sought in its partners.

Heart, or Intention

First and foremost, Tuggun had the firm *intention* to make life better for someone: his future self. Everyone wants a better life, but excessive hardship of a kind that is all too common leaves many individuals resigned to lives of poverty and powerlessness. Learned helplessness squashes aspirations, leading people in poor or oppressed circumstances to focus narrowly on the present to the detriment of their futures.[8] Tuggun, though, was not like this. "My parents never made it in life," he told me. "I was very determined to break that cycle of poverty." He had a strong intention to rise above the circumstances in which he found himself. He did what he needed to survive, but he also kept one eye on the hope of a better life. He avoided people who were likely to get him into trouble. And he kept looking for opportunities to climb out of his circumstances.

Taking care of one's future self is just one level of positive intention, though. Tuggun now fights for the rights of other disabled people in his country. Social causes are better served as individuals with a narrow present-orientation expand to a future-orientation; and as people extend their concern from themselves to others in what philosopher Peter Singer has called "expanding circles."[9] It's good to care for oneself, better to care for family and community, even better to care for country, and best to care for humanity as a whole.[10] The "evil" of tyrants and criminals often lies in their tiny circles of concern–they may have positive intentions for their present selves, and possibly for their future selves, but the concern doesn't extend to others. At the other extreme, saints have consistent positive intention for a large radius of sentient beings. The rest of us fall somewhere in between.

Changing intentions is hard to do, but it's the heart of social progress. The world's most meaningful social shifts–from slavery toward emancipation, from racism toward equality, from routine warfare toward routine peace, and from women-as-property toward gender parity–reflect ongoing mass changes in human intention.

Mind, or Discernment

Another aspect of Tuggun's rise is a canny *discernment.* Knowledge is one requirement for discernment, and I won't belabor its value. But discernment also requires the ability to make shrewd judgments about people and opportunities that go beyond bookish erudition.

Tuggun assessed his own situation lucidly, and he made choices that were appropriate for each moment. He realized that a good education had value. "My decision to accept the partial scholarship to study at Ashesi University was the best decision I have ever made," he said. He could tell which people would help him and which would bring trouble. He also exhibited social skills appropriate for interacting with the urban middle class, a kind of "cultural capital" that he'd picked up in part through observation.[11] And even without the rigors of modern psychology, Tuggun intuited that viewing situations as opportunities to learn was more worthwhile than dwelling on outcomes: "Unpleasant and undesirable consequences can be encountered but it is part of learning. Life is continued schooling."[12]

Most readers will take such sensibilities for granted, but they aren't always learned by people raised in less privileged communities, if only because of lack of exposure and practice.[13] Of the three pillars, discernment is perhaps the hardest to pin down, because it depends so much on context. We can never know whether a decision is optimal without knowing all of its downstream consequences. Was Tuggun right to have abandoned his college education? It worked out well for him, but it might not for another person in another context. Discernment–or prudence, judgment, practical wisdom, Greek *phronesis*–is not easy to teach or specify, yet we all know people who are sagacious decision-makers.[14]

Will, or Self-Control

Finally, Tuggun showed an incredible degree of self-control. I thought he was a determined student, but I only learned later just how determined. Tuggun revealed that he had been homeless while studying at Ashesi. "I slept on a bench under a tent at a public lorry park. The tent was not completely covered. There was no door. I had access to the bench only when the station closed for work at 11:00 p.m.," he said. "I had my bath at a public bath meant for passengers. I walked most of the time to and from campus, covering a distance of eight kilometers a day, and studying mostly on an empty stomach." He managed on a very tight budget, and he avoided temptations that could threaten his survival. Doors closed in his face, but he persevered. He worked hard to learn the math in my class, and if his emails over the years are any indication, he has also put effort into his writing. His story is an extended marathon of self-control.

Self-control allows us to follow through on what we intend or what we discern to be the best course of action.[15] It's one thing to yearn for the security of savings; it's another to muster the will to save. It's one thing to recognize the need for a vocational skill; it's another to expend time and effort to obtain it. It's one thing to know that collective action can overcome repression; it's another thing to risk imprisonment to organize.

How did Tuggun develop self-control? Commonsense notions about willpower are probably not far off–use it, or lose it; no pain, no gain. Psychologist Roy Baumeister finds self-control to be like a muscle. In the short term, it can be exhausted by intensive use, but with regular exercise, capacity can be increased in the long term. Tuggun might have gained self-control by practicing the childhood habits ingrained in him by his parents, by raising his younger siblings after the death of his parents, by managing the obstacles his physical disability presented him with, and by going through the academic hoops at Ashesi.

Reviving Dead Sages

Combined, the capacity for intention, discernment, and self-control–or again, heart, mind, and will–might be called virtue, character, maturity, emotional intelligence, *sophia*, or wisdom.[16] Unfortunately, these terms are all loaded. They're fraught with religious, political, and philosophical dogma that fails to distinguish between pure expedience and moral righteousness.

My hope in the rest of the book is to show how heart, mind, and will–not just in individuals, but in societies–are the essence of social change for purely practical reasons that transcend creed, culture, and politics. To do so, I need vocabulary without baggage. At the risk of introducing yet another phrase that time could taint, I'll use the term *intrinsic growth* to describe progress in intention, discernment, and self-control. These are attributes fostered within a person or a society, in contrast to external, technocratic, packaged interventions. Intrinsic growth is also allied with the notions of intrinsic motivation–the motivation a person feels from within, rather than that inspired by external rewards or punishment–and intrinsic learning, to which we'll return in the next chapter.

Occasionally, I'll use the words "wisdom" and "virtue" interchangeably with intrinsic growth, but one reason those words aren't ideal is that they call to mind either old gray-haired folk or demure young virgins. Intrinsic growth, though, isn't about age or sex–it's about improving intention, discernment, and self-control. The point is not to turn individuals into long-bearded gurus, but to nudge everyone toward incrementally greater intrinsic growth.[17]

In Part 1 we saw how no amount of educational technology makes up for a lack of focused students, caring parents, good teachers, and capable administrators. So, what is it among the latter that matters? Focused students have the intention to learn; the discernment to listen (selectively?) to supervising adults; and the self-control to study. Caring parents intend to nurture self-sufficient children, recognize good schooling, and intervene just enough to hold educators accountable. Good teachers have their students' best interests at heart, make hundreds of small judgments every day to enhance learning, and do all this without losing their cool in a potentially adversarial classroom. And capable principals intend, discern, and act to manage schools well.

We also saw in Part 1 how democracy requires much more than Facebook revolutions and ballot boxes. It demands active citizens, effective bureaucrats, and enlightened leaders.

To take another example, it's increasingly clear that stabilizing or reversing climate change will require mass intrinsic growth. As individuals, we need the intention to leave a sustainable world for our descendants, the discernment to recognize the urgency of the situation, and the self-control to curb consumption. Our corporations need intention, discernment, and self-control to place long-term value above short-term profit. And our political leaders need the same intention and discernment, as well as the self-control to resist special interests and careerism.

None of this is easy to package, and all of it requires intrinsic growth.

I don't mean to suggest that intrinsic growth is the entirety of societal progress. Laws, vaccines, schools, laptops, markets, agronomy, manufacturing technology, clean energy, ballot boxes, economic policy, transportation infrastructure, and government institutions are all pieces of the puzzle. These packaged interventions have great positive impact if they are implemented well.

But "implemented well" is the rub. Perhaps the greatest implicit misconception about social causes is that the right cookie cutter is what's important: If we could just identify the right mold, we could mass-produce it, ship it off, and cut dough wherever cookies are needed. Some cookie cutters might be better than others, but the tools aren't what make a good cookie. It's the heart, mind, and will of the chef that matters. Similarly, television can be a part of a good education, but only with wise teachers directing their use.[18] Microcredit can alleviate poverty, but only with wise institutions melding efficiency and compassion.[19] Elections can yield a responsive government, but only with wise citizens ready to impose checks and balances. When positive social change happens, it is because there is a base of intrinsic growth holding up the endeavor. It's not that packaged interventions aren't important, but that intrinsic growth is their ultimate controllable cause. If we focus on intrinsic growth, the rest will take care of itself.

Intrinsic growth is not a new concept; it is old ideas considered in a new light. Every traditional virtue can be described in terms of the three pillars: courage overcomes fear with self-control at well-discerned moments for the sake of someone other than one's present self; temperance discerns against short-term reward and applies self-control for long-term advantage; justice and compassion are expressions of good intention toward others; prudence is self-control discerningly applied; humility discerns where confidence becomes overconfidence and asserts control over pride. Recently popular virtues such as grit and resilience are similar recombinations of heart, mind, and will. Cross-cultural analyses show that these and other virtues are valued throughout the world, even if their relative emphasis varies.[20]

When you ask people who they believe to be civilization's wisest people, they nominate people like Socrates, Mahatma Gandhi, Mother Teresa, Benjamin Franklin, Nelson Mandela, Daw Aung Sang Suu Kyi, and so on.[21] Notably, the lists typically exclude the likes of Mozart and Steve Jobs, even if the latter might have been wise in limited domains. Intelligence, talent, and brilliance aren't the same as heart, mind, and will, although some IQ might be needed for good discernment. Gandhi went on hunger strikes to free India from British rule, and Mandela emerged from twenty-seven years in prison only to seek reconciliation with his captors. More than smarts, what these actions required was saintly intention and extraordinary self-control. Gandhi and Mandela cared for a wide circle of people beyond their familial, ethnic, and national affiliations, and they pushed through granite obstacles with superhuman will.[22]

Finally, why three pillars, and why these three? It's because they're the minimal building blocks for all positive human action.[23] The wisdom necessary for long-term health provides an example: For optimal health, it's necessary to want it (good intention toward your future self), to understand the benefits of nutrition and physical activity (discernment), and then to eat well and exercise (self-control). If you're missing any one of these three components, better health is less likely to follow. For example, a workaholic nutrition scientist may know what a good diet is (discernment), and have great willpower (self-control), but still be cavalier about her health, because she's too focused on her research (bad intention). A superstitious health fanatic may want good health (intention), and spend effort to do what he believes to be healthful (self-control), but be misguided by unsound medical advice (bad discernment). And a smart, well-intentioned couch potato may want health in the abstract (intention), and know well what's good for her (discernment), but still be too lazy to get off the sofa (bad self-control).[24]

Group Intrinsic Growth

The lessons of intrinsic growth apply not just to individuals but also to groups. Public health is a good example. More than other social-cause disciplines, it has hosted a vigorous debate about the technology-and-intrinsic-growth balance. Modern health care can be seen as a march of amazing technologies, and we ought to celebrate the parade. But in public health, where the question isn't just about biology but also the social underpinnings of mass health, many experts emphasize the human institutions that deliver health care–what they call "health systems."[25] As a result, more than in other tech-heavy fields, public-health practitioners make conscious efforts to cultivate intrinsic growth in organizations and societies.

The International Training and Education Center for Health at the University of Washington, or I-TECH, offers an example. It supports "the development of a skilled health work force and well-organized national health delivery systems" around the world.[26] Its founding director is a global health professor by the name of Ann Downer, a woman whose sharp, steely temperament resides beneath a warm, motherly glow.

I-TECH was founded in 2002, but Downer credits I-TECH's rise to funding from PEPFAR, the President's Emergency Plan for AIDS Relief. And she credits PEPFAR's strategy–which came as a pleasant surprise to many in public health–to its founding director, Mark Dybul. Dybul is an infectious disease expert who understood that dumping antiretroviral drugs in countries plagued by AIDS wouldn't accomplish much. Health-care capacities in affected countries needed to be built up. PEPFAR intended from the outset to help government health ministries become self-sufficient in combating HIV/AIDS, and I-TECH was funded because it focuses on strengthening health systems.

I-TECH operates at various levels of a health system, but all of its work comes down to training and organizational development. It teaches frontline health-care providers, instructs lab workers in quality control, offers leadership training to health ministry officials, develops standardized curricula for health-care education programs, and trains local instructors to do all of these things on their own. "We do less with frontline health workers than we used to," Downer said. "But whether frontline or up the chain of command, our goal is always to help health systems become more effective and self-sustaining for the ultimate benefit of patients."

There's little glitter in I-TECH's work. Downer was matter of fact when I asked her about the specifics of teaching in Tanzania. "Here in the United States, I think we take a lot for granted," she told me. "We minimize the powerful effect of a good teacher using basic, low-tech, and replicable teaching methods–things like discussion and case studies to help learners develop a body of knowledge and critical thinking skills. These same methods can have a profound impact on learning when applied in countries struggling to provide basic education in an era of high-tech temptation. It gives new meaning to 'back to basics.'"

Basic efforts, though, can have lasting impact. In the decade since its founding, I-TECH has provided training to over 180,000 health-care professionals, contributing to an estimated 1.2 million lives saved during PEPFAR's first four years.[27] Every life saved, of course, has ongoing impact through families and economic productivity. But these numbers still underestimate the eventual impact of a stronger health-care system beyond HIV/AIDS treatment.[28] Researchers cite improvements in general health care where PEPFAR works, including more reliable blood supply and more households treating their drinking water.[29] And the training lasts throughout a health-care worker's professional life. Much of it will be passed on to those they manage and mentor.

It's impossible to tally all of the downstream benefit. Who knows how many great-grandchildren will have better health because an ancestor started boiling water at home at the suggestion of a nurse who learned from an I-TECH curriculum? And it's important to acknowledge that it's impossible. By doing so, we shift from mere cost-benefit analysis to a judgment involving less measurable desiderata: Should we simply provide as many packaged interventions as we can, or should we nurture institutions that can fund, maintain, and implement packaged interventions on their own?

What organizations such as I-TECH do amounts to building the heart, mind, and will of national health-care systems.[30] The concept of intrinsic growth applies not just to individuals but to groups and societies as well. Of course, how these traits are formed and expressed differs for groups as compared to individuals. Group heart, mind, and will are the result of many individuals combining their intentions, discernments, and self-control through organizational structures and messy but unavoidable politics. Social qualities such as trust, which have no real meaning for a single person, suddenly take on great importance as they mediate group interactions. Nevertheless, groups have intentions, groups discern among options, and groups act with varying degrees of self-control. Group intrinsic growth is a complex aggregation of individual intrinsic growth with complicated social factors mixed in.[31]

Downer noted, however, that their work was dependent on a preexisting foundation: "For I-TECH's work to be effective in the public health arena, the public education system must also be functional," she explained. "When all systems work together, you are then left with the most important elements–individuals' commitment to their jobs and their nation and their ability to set goals and get things done." While these were traits that I-TECH could strengthen to some degree, it is difficult to foster them from scratch.

The same could be said about Ashesi University. Despite its accomplishments, the school can't take full credit for the achievements of its alumni. Like other good universities, Ashesi is selective about whom it admits. That means many students arrive with strong heart, mind, and will to begin with.

So both I-TECH and Ashesi force us to ask another set of questions: How do you establish the foundation required for people to benefit from an I-TECH or an Ashesi? And is it possible to establish that foundation in the starkest of circumstances? These questions are answered by the miracle that is Shanti Bhavan.

A Haven of Peace and Learning

When she was in the eleventh grade, Tara Sreenivasa wrote that she liked math and computers. Thanks to a terrific teacher, she especially enjoyed accounting. She also took guitar lessons, and in the afternoons she could be found playing badminton or basketball with her friends. She explained her future plans confidently: "I want to get into . . . business and earn lots of money. Someday, I would like to start a company of my own and open an old age home in remembrance of my grandmother . . . because she has taught me to be independent."[32] In 2009, Sreenivasa passed her state's standardized secondary-school examinations and went to college. Upon graduation, she became an accountant.

Sreenivasa might not seem too out of the ordinary, but she's extraordinary when you consider her background. She comes from Kundhukotte, an extremely poor agrarian village in South India. The Indian government designates her family "Backward Class," a label applied to castes that are formally considered "socially and educationally backward."[33] Sreenivasa's mother had no formal schooling and cannot read or write. She used to work as a cook in a government school, where she earned $120 a year. Sreenivasa's father is unemployed and mostly absent. As for her grandmother, Sreenivasa thinks of her as independent because, until recently taking ill, she provided for herself by doing menial farm labor on other people's land for wages of 40 cents a day.

Meanwhile, in a village a couple of hours away from Kundhukotte, a woman roughly Sreenivasa's age illustrates what life could have been for Sreenivasa. Kavitha's life began much like Sreenivasa's, but Kavitha stopped going to school in the eighth grade. She had chores to do, and the four-kilometer trek to her secondary school took too much time. In any case, she was learning little by the end of primary school. Her school system stuck to a rigid curriculum regardless of student comprehension; by fifth grade, she and most of her classmates had fallen hopelessly behind. At the age of fourteen, Kavitha was married to her uncle, fifteen years her senior, to whom she had been promised since childhood. He is the groundskeeper for some local government offices. It's a steady job that pays a couple of hundred dollars a year, but it offers no prospects. He doesn't believe Kavitha should leave the house except to buy food. When I met her in 2009, she was eighteen and pregnant with her second child.

Sreenivasa has a brighter future thanks to Shanti Bhavan, a boarding school where children of low-income, low-caste Indian families are provided a first-class, donor-funded education. With her parents' consent and encouragement, the school adopted Sreenivasa in all but name at the age of four.[34]

Shanti Bhavan occupies forty acres in the middle of a stretch of agricultural land in the South Indian state of Tamil Nadu. For miles around, there is little else but small, partitioned fields in various states of cultivation. Baliganapalli, the nearest village, comprises perhaps a hundred mud, brick, and concrete homes. Its roads are unpaved, and the biggest building is a small government primary school.

In contrast to Baliganapalli, Shanti Bhavan is a lush oasis. Abraham George, the school's visionary founder and main benefactor, translates Shanti Bhavan as "Haven of Peace." He felt that his students should have facilities equal to those enjoyed by their upper-middle-class counterparts.[35] Instead of the hard, dry dirt surface of so many Indian schoolyards, well-trimmed grass covers the grounds at Shanti Bhavan. Carefully tended bushes and flowers line the walkways. In May, the bright orange flowers of "flame of the forest" *gulmohar* trees light up the driveway like fireworks. The school building is a pastel pink two-story structure with a large courtyard that all the classrooms face.

Lalitha Law was Shanti Bhavan's headmaster for its first twelve years. "We have to start with basics that middle-class schools can take for granted," she told me. Few of the students have seen a book before entering the school. No one comes in knowing the alphabet in any language. But there are even more basic challenges. "When they arrive, few of the children have modern health practices," Law said. "Some are not used to regular bathing. Most do not have a habit of brushing their teeth." She hinted that many of them go to the bathroom right on the school lawn.

Shanti Bhavan thus has a staff of "aunties," who raise new entrants as they would their own children. Sreenivasa remembers her first few days: "I learnt how to hold a spoon and how to polish my shoes. I saw a toilet for the first time in my life. I was taught how to brush my teeth, wear my shoes, to eat with a spoon, to hold a pencil and a book, and many other things." By the time Shanti Bhavan children formally enter the first grade at age five or six, they are not quite at the level of their middle-class peers, but they're closer.

The remaining gaps, however, are closed over the next twelve years as the students receive an education on par with that provided by first-rate Indian private schools. The dedicated faculty are supplemented by a stream of volunteers, mostly from abroad, who focus on extracurricular activities. Students play the piano, put on plays, and compete in soccer tournaments. (Incidentally, the school is not particularly high-tech. It has one classroom for an elective computer class outfitted with PCs, none of which are connected to the Internet.)

By secondary school, Shanti Bhavan students are indistinguishable from children of well-educated, upper-middle-class Indian households. Every one of their graduates so far has gone on to a good university. In 2012, the first batch of Shanti Bhavan students finished college. They now work at various multinational companies, such as Goldman Sachs, iGate, and Mercedes Benz. Sreenivasa herself works at Ernst & Young. Without Shanti Bhavan, these same children would have been lucky to be literate. They would have likely ended up in wage labor, marginal farming, or other work below the poverty line.

I have visited many other alternative schools in India, but none of them were anything like Shanti Bhavan. Schools that spend two or three times the government budget per student can increase learning here and there, but only the occasional student will see a radical change in his or her future. At Shanti Bhavan, the transformation is total. The school's eleventh-grade students once challenged my colleagues and me–PhDs with years of public-speaking experience–to a debate. They trounced us with their poise and their well-constructed logic, all while regaling us with quotations from Shakespeare. And though sending every tot to boarding school may not be a viable public policy, Shanti Bhavan still stands as a beacon that demonstrates the dramatic effect of a good all-around education–one that stresses academics, extracurriculars, character, and cultural capital. In one generation, Shanti Bhavan ejects its students from poverty.

The Real Value of Formal Education

Focusing on financial poverty, however, encourages a narrowly technocratic view. It's tempting to see Ashesi University and Shanti Bhavan as producers of economically productive human capital. Much of their success is visible in the job placements of graduates. As World Bank economist Harry A. Patrinos and his colleague George Psacharopoulos noted, "it is established beyond any reasonable doubt that there are tangible and measurable returns to investment in education." Based on data from a range of countries, they estimated the economic rate of return to nationwide education programs to be roughly 10 percent.[36]

But Nelson Mandela once said that "education is the most powerful weapon we can use to change the world," and he was certainly not just talking about economic change.[37] In fact, education's benefits go well beyond economic productivity. Here, for example, is Patrinos's own catalog of the benefits of girls' education:

> A year of schooling for girls reduces infant mortality by 5 to 10 percent. Children of mothers with five years of primary education are 40 percent more likely to live beyond age 5. When the proportion of women with secondary schooling doubles, the fertility rate is reduced from 5.3 to 3.9 children per woman. Providing girls with an extra year of schooling increases their wages by 10 to 20 percent. There is evidence of more productive farming methods attributable to increased female schooling and a 43 percent decline in malnutrition. It has also

been shown that educating women has a greater impact on children's schooling than educating men. In Brazil women's resources have 20 times more impact than men's resources on child health. Young rural Ugandans with secondary schooling are three times less likely to be HIV positive. In India women with formal schooling are more likely to resist violence. In Bangladesh educated women are three times more likely to participate in political meetings.[38]

Statistics such as these lead some development specialists to think of girls' education as the closest thing to a silver bullet in the fight against global poverty.[39]

What's not evident in these figures is why education has such wide-ranging impact. Patrinos links education to productive farming, but what does formal schooling have to do with agriculture? Modern curriculums don't teach it. Why does a girl's primary schooling affect the survival rate of her children? She's not learning about neonatal care in third grade.

As highly regarded as education is,[40] it is not always appreciated for the right reasons. Part of the problem is that we think of education as being about grammar and multiplication tables, about names and dates and cognitive skills. But while knowledge is essential for a productive life, much more is necessary.

Often overlooked is the deeper transformation that good education can bring about in heart, mind, and will. It's not just that Sreenivasa is more knowledgeable or "better educated" than Kavitha, it's that Sreenivasa has an aspirational outlook, a belief in herself and in her ability to learn, intrinsic motivation toward many interests, and a desire to contribute to causes beyond her own–all traits that are largely neglected in policy circles.

Fortunately, hard evidence for education's less tangible value is slowly emerging. In one noteworthy study, economists Pamela Jakiela, Ted Miguel, and Vera te Velde show a striking change in attitude among Kenyan teenage girls based on just two years of formal schooling.

Among a group of over 1,800 girls, half were randomly chosen to receive a scholarship. And in the way of behavioral economists, Jakiela and her colleagues had the two groups play games–specifically, variations of the game of Dictator.

In its standard version, Dictator involves two players, one of whom is given a fixed amount of cash. The sole action of the game occurs when that person is offered the chance to give away all, some, or none of it as she pleases to the other player. In its simplicity and ubiquity, Dictator is the tic-tac-toe of behavioral economics. Repeated experiments show that most people in the dictator role prefer to keep a majority of the cash for themselves but still give a portion to the other player. In other words, as dictators, most people aren't self-sacrificing saints, but neither are they total misers. The question, though, is how much they keep.

In this case, the economists devised four variations of the game. In versions 1A and 1B, the cash was given to the dictator as usual. In versions 2A and 2B, the cash was given to the other player, and the dictator could choose to take as much as she wanted. In A games, the amount of the pot was determined by a dice roll by the player initially receiving the cash. In B games, the pot was decided by a physical exercise: The harder the player worked, the more total cash was disbursed. That is, in the A games, the pot was decided by luck; in the B games, the pot was decided by the effort of one of the players.

The main result was that, as dictators, the girls who attended school kept more when they put in the effort, thus rewarding themselves for their work. They also gave more when the other player put in the effort, thus respecting the others' exertions. The effect was most pronounced in students whose grades had improved in the preceding two years. The researchers suggested that the students internalized the experience of being rewarded for effort at school and brought that ethic to other parts of their lives. Just two years of formal schooling appeared to instill a healthy valuation of effort over luck.

Valuing effort instead of luck shows both good intention and discernment–greater heart and mind. It not only affirms an individual's capacity to reach goals, but also strengthens social norms to reward effort. On the whole, people are more likely to reach their goals if they believe that greater effort matters, and societies are more likely to prosper if they reward effort. Notably, it doesn't matter whether real-world outcomes are primarily based on luck or effort. If effort is even 1 percent of the cause, it's still good for people to believe in it. Luck averages out, but the effects of that 1 percent will accumulate over time like compound interest.[41] Indeed, as early as 1966, sociologist James Coleman suspected a related cause for social disparities in a study commissioned by the US government. One of the key differences between children of different socioeconomic classes in America was the degree to which they believed in luck or effort.[42] The children of better-off, more educated parents valued effort.

The study by Jakiela et al. demonstrates that education is much more than a transfer of knowledge. Education builds heart, mind, and will, too. Something about schooling is more conducive to intrinsic growth than is, say, twelve hours of mindless menial work in rice fields or in a sweatshop. It's not that rice fields and factories can't be educational: Some farming and industrial communities appear to transmit great wisdom across generations.[43] And some forms of alternative education use the farm or workshop as the classroom.[44] But these environments are more often the site of child labor or child neglect–even when benign, they rarely engage kids with continual learning opportunities.

In contrast, with an effective education, there are repeated chances to learn, "I can do this!" This is true even in the much-maligned brand of education that is rote learning. Japan, for example, has a national education system based largely on cycles of listen-study-test-pass, but few would argue that Japan's educational system is failing.[45] The country has very high literacy rates, a long life expectancy, and the third-largest economy in the world despite a prolonged recession. Similarly, elite students from China and India outperform average students in the United States even though they're raised on rote learning. The base of optimistic intentions, keen discernment, and greater self-control that a good rote education develops is far better than no education, poor education, or an ambitious educational program run badly.

None of which is to say that rote learning is the pinnacle. Hardly. The education that the world's most privileged children enjoy is often tailored to foster their unique talents. Great teachers prod, guide, motivate, and inspire. A good school encourages collaboration and creativity.

And well-off parents provide their children with a range of enrichment activities that go well beyond reading, writing, and arithmetic.[46]

In fact, a good education is one of the few things that statistics-minded technocrats acknowledge exist, even as they are unable to quantify its causes. Small things can matter, and too many of them are lost in efforts to cut costs or to standardize achievement. For example, good schools often bring in outside speakers to expose students to a variety of professions. Though this in no way improves test scores, it can help students develop life goals. Shanti Bhavan hosts many visitors; its students aspire to be doctors, engineers, scientists, journalists, environmental activists, and Broadway singers. A disproportionate number want to be astronauts, because of a memorable visit by NASA astronaut Sandra Magnus. Meanwhile, surveys of Indian government school children reveal that their hopes are limited to government jobs–those are the only stable jobs that they've heard of.[47] As Plutarch wrote, "the mind is not a vessel that needs filling, but wood that needs igniting."[48]

True Sustainability

If good education is effective, it is neither quick nor cheap. George exhausted his fortune underwriting Shanti Bhavan for its first fifteen years, and recently the school has had to cut back on enrollment as it builds its fundraising capacity. Besides, even if Shanti Bhavan itself were to endure, what about the 250 million other children in India who aren't in good schools? It's unlikely there are a million Abraham Georges to fund their education. Is Shanti Bhavan just a nice one-off story without a generalizable lesson? Is its model of education sustainable?

Policymakers and large foundations often speak of catalyzing a transformation with a finite injection of cash. They talk as if social change is a cascading domino rally just waiting for someone to come along and knock over the first piece. When they seek "sustainability," what they really want is to take immortal credit for having made a limited one-time donation. As a result, meaningful expenditures on education or capacity building are often seen as unsustainable, because they incur high costs and require someone to keep paying for them year after year. (Here again is C. K. Prahalad, who, as you may remember, groaned that "charity might feel good, but it rarely solves the problem in a scalable and sustainable fashion.")[49]

This is, however, a counterproductive view of sustainability. Shanti Bhavan, for example, is a decidedly unsustainable endeavor by the standards of Prahalad. It costs the school about $1,500 per year per student, and these funds are not recovered from the students' families–they certainly couldn't afford it. Nor does the school's model seem expandable under any large-scale government program, at least in India. The $1,500 per year dwarfs what the Indian government spends on each of its public school students, which is less than $250 a year.[50] The fate of Shanti Bhavan, then, is dependent on charity. But it would be a great mistake to believe that Shanti Bhavan's impact isn't sustainable.

Sreenivasa herself is a towering monument to sustainability. As a Shanti Bhavan graduate, Christ College alumna, and Ernst & Young employee, Sreenivasa is firmly embedded in upper-middle-class India. Even if she never makes more than $5,000 per year–a low estimate, given her skills and experience–she would have an excellent standard of living compared to the hundreds of millions of Indians living on just a few hundred dollars a year. Barring debilitating setbacks, Sreenivasa will never know the kinds of struggles Kavitha, whose household income probably will never exceed $1,500 per year, will experience. And if they are typical of their peers, Sreenivasa will be happier than Kavitha.[51] So Sreenivasa's story is a triumph even if Shanti Bhavan were to shut its doors tomorrow. As avuncular figures say all over the world, "education is the one thing that no one can take away from you." Shanti Bhavan's effect on its students ratchets.

But there's more. Sreenivasa will command jobs that provide good health care for her family, while Kavitha will be lucky to have a competent rural health-care worker to call on. Sreenivasa will be able to send any children she may have to good schools. Kavitha will make do with understaffed government schools or marginally better private schools (where fees are about $1 to $2 a month). Sreenivasa will pay taxes and contribute to her parents' welfare. Kavitha may have to beg from neighbors to survive dry seasons.

None of this is to say that Kavitha's life is doomed, or that outcomes are determined wholly by internal traits. Kavitha might join a local self-help group and become a community leader. Or she might inherit land from a rich relative. And Sreenivasa is not immune to hardship. She may be laid off or lose money in a stock-market crash. Both women will experience ups and downs. Still, the range of likely outcomes for Sreenivasa and Kavitha is bracketed and persistent. Sreenivasa is in a better position to live well and to contribute to the well-being of her children and their offspring.

And if that's not enough, yet another effect of Sreenivasa's education is felt by her home community. Some critics of Shanti Bhavan question the stark sociocultural rifts caused within families when one child is taken to what might as well be another planet. I once shared these concerns.[52] My reservations evaporated, however, when I visited Shanti Bhavan's campus and met their students. The school takes pains to retain the children's ties to their families through careful counseling and two home leaves each year. Students are open and honest about the challenges of staying close to family. Some have navigated complex situations such as sexual abuse, which can be routine at home.[53] Most acknowledge the gaping social, cultural, and linguistic canyon between themselves and their families and have become accustomed to mediating any differences.

Despite the challenges, the students are well adjusted, and their families report joy and amazement at their progress. Students are often treated like celebrities when they return home, and they inspire neighbors to seek out the best schooling they can afford for their own children.[54] Because Sreenivasa's mother saw the effect of Shanti Bhavan on her daughter, she did everything she could to keep Sreenivasa's sister Gayathri in school. Gayathri now attends a government college, and her own future is much brighter than it could have been.

Lastly, Sreenivasa also bears a kernel of the school's financial sustainability in a way that might even have made Prahalad proud. In 2013, I took Sreenivasa and some of her friends out for lunch. All of them were from Shanti Bhavan's pioneering classes, and the first to start working.

But even though they had only been in their jobs for a few months, they had already begun contributing a portion of every paycheck to Shanti Bhavan. As more students graduate, the balance of supporters to students will shift until, at some point, alumni could fund the whole student body, and perhaps expansion to boot.

All of that–what Sreenivasa keeps for herself and propagates to future generations–comes without another penny from Shanti Bhavan, its founder Abraham George, or any other benefactor. That's a lot of sustainability.

. . . And Scalability

Unlike Shanti Bhavan, Ashesi University is already financially viable, as its operating costs are covered by an admittedly high tuition. (Donors contribute to scholarships, capital campaigns, and endowments.) But Ashesi also provides a hint toward large-scale impact. University president Awuah argues that if today's African college graduates–just 5 percent of every age group–could be persuaded to work toward Africa's future, the continent would be transformed in twenty years when that 5 percent would inevitably assume leadership.[55]

Twenty years might seem a long time to wait. Those feeling the burden of the world's thousands of corrupt politicians, millions of starving children, and hundreds of millions of subsistence workers might say that we can't afford to spend two decades nurturing individuals when people are suffering right now. But that is less an argument for a particular course of action than a lament for unavoidable tragic tradeoffs.

The best insurance against this suffering is a country's own development. But taking a nation from a dollar-a-day income to $11,670 a year per capita, the US poverty level for a single-person household, would require about four decades at a sustained breakneck growth rate of 10 percent a year.[56] In comparison, it only takes twenty years to raise a new generation. Speed is relative. China and the East Asian tigers pulled themselves up in a matter of decades, thanks in great part to widespread investments in high-quality universal education. With Ashesi, Awuah is helping to lay a new foundation for Africa.

Progress can't be taken for granted, but even small efforts to raise intrinsic growth tend to be self-sustaining. And big efforts, such as the one that raised Sreenivasa, are truly transformational.

CHAPTER 8

Hierarchy of Aspirations

The Evolution of Intrinsic Motivation

March 1, 2012, began like any other day for Regina Agyare, a software engineer at Fidelity Bank in Ghana. She woke up, got dressed, and had breakfast. She drove to work. She logged onto her computer and checked her emails. But that day wasn't like any other day. That day, she quit.

She had tried to quit once before. The company wanted to keep her, though, and "they countered with a promotion, a raise, and other incentives," she told me. That first time, Agyare decided to stay.

But not on March 1. "That day, too, the bank tried to convince me to stay," Agyare said. "My manager suggested that I stay at least until the end of the month when the bank gave out employee bonuses." It was tempting, and, again, she reconsidered. But not for long. Something in her just knew it was time to go. By the afternoon, "I packed up my office and left for good."

"I didn't have another job lined up, and I didn't have any grand plans," Agyare said. But she had a rough idea what she wanted to do. Two weeks later, she started Soronko. Soronko means "unique" in the Ghanaian language of Twi, and it lives up to its name. It's actually a pair of entities: a for-profit business called Soronko Solutions and the nonprofit Soronko Foundation. "Soronko Solutions provides software development services to small and medium businesses, which are underserved in Ghana even though there are so many of them," Agyare explained. From the revenue she makes through Soronko Solutions, she applies an astounding 80 percent to fund Soronko Foundation, her true passion. Soronko Foundation teaches technology skills to Ghanaian youths. Its Growing STEMS program provides rural kids with supplementary science and technology classes. "We recently started another program called Tech Needs Girls," Agyare told me. "We organize women engineers to mentor girls from urban slums in computer programming and entrepreneurial skills."

Agyare's work has not gone unnoticed. She has been crowned with laurels by the World Economic Foundation, the Aspen Institute, and Hillary Clinton's Vital Voices Fellowship. Facebook executive Sheryl Sandberg wrote about Agyare in *Lean In for Graduates*.[1] And in 2014 Agyare was selected for the Young African Leaders Initiative begun by President Barack Obama.

In previous chapters, I've discussed the best ways to deploy packaged interventions and to nurture the intrinsic growth needed to implement them. But social causes need more than that. They also need people who, leading or following, put in time and resources for the sake of others. Just a few years ago, Regina Agyare appeared to be just like the hundreds of millions of other comfortable adults in the world's middle and upper classes. She was doing well for herself (which is important!), but her positive impact on others was limited. Then she made a transition.

Advanced Intrinsic Growth

What differentiates the effective social activist from everyone else? And what causes the difference?

I know Agyare well because she was the top student in my calculus class at Ashesi University. Then as now, she had the bearing of a queen, confident and graceful. She liked doing things her own way and at her own tempo. (She also once threw a pie in my face. I had assigned a tedious problem set, and I wanted to give the students a chance at payback. She rushed to volunteer as the class's angel of revenge.)

Back then, Agyare had little interest either in starting her own firm or in mentoring groups of Ghanaian girls. She was focused on securing her own future. "One of the classes I took at Ashesi was about entrepreneurship," she said. "I told the lecturer that I didn't see myself starting a company. At the time, I just wanted a dependable job." She studied computer science and expected to become a programmer.

But while Ashesi offers majors in technical subjects, its mission is much broader; the school's motto is "Scholarship, leadership, and citizenship." Agyare remembers that the school expected students to become "a new generation of ethical leaders," whether in business, academia, government, or other fields.

The administration tried to model this behavior through a standing policy of never paying bribes, even though that meant that some things slowed to a snail's crawl in the Augean muck of local bureaucracy. Extracurricular service activities were encouraged. In the year I taught there, students raised funds for a local charity that worked with blind children. And an honor code the students proudly followed held them to a high standard of integrity. During the university's accreditation process, the government questioned whether students could be trusted to police themselves and recommended stricter oversight. The students rebelled with a passionate and ultimately successful appeal. They argued, "If we cannot maintain a culture of honesty as students, then how can we be expected to do it when we grow up to become the nation's leaders?"

The seeds that Ashesi planted in its students bloomed in Agyare. "We'd been taught that we could change the world," she said, and she wanted to do her part. But how? "When I graduated, a good friend of mine started a business right away. But I wasn't sure that I could do the same–I had no capital, no experience, no network." So, for seven years, she worked a series of corporate jobs. She bided her time until she was ready.

When I asked Agyare why she finally left Fidelity Bank, she had many reasons. "My first manager was terrific–he understood technology, gave me credit for my work, and let me set my own milestones," she said. "But he left, and let's just say my new manager did less of those things." There had also been a series of reorganizations at the firm, the result of which was to leave Agyare with pettier colleagues and fewer interesting opportunities. "The straw that broke the camel's back was when the bank started outsourcing technology projects to foreign companies," she said. "When I first joined the bank, I liked it because they promoted it as a local bank built and run by Ghanaians."

Meanwhile, her desire to strike out on her own was gradually growing. Slowly, she gained experience. Slowly, she gained confidence. And slowly, she built up a supportive network. About March 1, she said, "Even that day when I woke up, I didn't know I was going to resign."

In the years before she quit, Agyare wanted to ensure a secure livelihood. She had positive intention for her future self. She had a solid base of knowledge and the discernment to land good jobs. She had the self-control to see her goals through. In a nutshell, she had enough heart, mind, and will to secure personal well-being and success.

But by the time she started Soronko, Agyare's intentions had expanded. She was no longer content to serve herself and her employer; she wanted to contribute to the betterment of children who lacked her advantages. She also had greater discernment and self-control: "I was just more confident and more experienced. I had skills and I had passion. And I remembered what I did at Ashesi, when I held leadership roles like being president of a social club, vice president of a women's group, and a peer educator about HIV/AIDS. By the time that I left, I just knew that I could do something on my own."

So the key difference between Agyare before and after her bank job is more heart, mind, and will, more intrinsic growth. Of course, there were other factors. After seven years of work, Agyare had some savings, making it easier for her to leave behind a steady paycheck. She also had a stronger network to call upon for support. But if Agyare had had those things at the start of her career, would they have been enough? She thinks not. "I didn't see myself as an entrepreneur," she said. "It required too much initiative."

Agyare's case suggests that with more heart, mind, and will–if you take intrinsic growth further and further–you reach a tipping point of sorts. A point where you become a net contributor to social causes. Intrinsic growth is both the basis for those suffering from social problems to rise above them and the reason why those who are secure in their own well-being contribute on behalf of others. The difference is one of degree: a larger radius of intention, keener discernment, and greater self-control. In other words, more intrinsic growth is the root controllable cause of all positive social change, whether you start poor or rich, oppressed or oppressing, powerless or powerful.

Changing People

Technocrats–especially the economists among them–say that people respond to incentives, and by incentives, they usually mean money. The standard "rational choice model" of economics postulates that everyone works selfishly to maximize her or his own utility, most often measured, again, in dollars.[2]

Of course, for many people, money is not the only incentive. It's often not even the primary one. Economists themselves offer plenty of counterexamples. Here are a set of rational people who are the world's experts on money. If their main goal were to maximize financial utility, they would all chase after the best-paying jobs on Wall Street. Some economists do just that, but there are plenty of others who become professors, policymakers, and journalists who aren't achieving their full earning potential.

What's more, there are economists who've held previous jobs in banking and finance and who have quit to pursue less lucrative careers. I've run into a few people like that, and the reasons they gave for switching include: "I got sick of the rat race, although I was a well-paid rat"; "I wanted to do something that was more intellectually rewarding"; "Time with my family became more important"; "I wanted to be my own boss"; and "I was looking for something with more meaning."

These responses betray two radical deviations from the dominant economics. First, none of the transitions were about earning more money. They were about family, autonomy, recognition, intellectual reward, and social impact. Money was still important–few took on work without pay, and many of them could afford to change jobs because they had a fluffy financial cushion. But the desire for wealth was satiated at some point, and other desires took over.

Second, people changed. Preferences evolved. Human nature isn't fixed–people's motivations change over time. Some who leave lucrative jobs take a cut in pay or give up a paycheck altogether. A few even begin giving away the wealth they've accumulated. Nor is this kind of change limited only to economists. Agyare is a standout example, but, as we'll see, she's hardly alone.

The Role of Aspirations

Whether it's economists or engineers, or, for that matter, farmers or factory workers, the cause of a voluntary life change is often a change in aspiration. In Agyare's case, we see a very clear shift from wanting a solid, dependable job to being her own boss and making a larger societal contribution.

Aspirations are potent forces, and it's a cliché to encourage their pursuit. But some truth must underlie this mother of all commencement-speech exhortations. In fact, there are at least four reasons why aspirations are so meaningful.

First, aspirations challenge a person to aim for something better, "for something above one," as the *Oxford English Dictionary* puts it.[3] Throughout her adult life, Agyare's aspirational concerns were for her own future welfare and then for the welfare of others. For some time now, I have been asking people I meet what they would like to change about themselves or their lives over the next five years. In Kenya, I inserted the question into a sample survey of 2,000 respondents who cut across all walks of life. So far, every single response has been positive. They fall into a dozen broad categories. They want to fulfill basic needs, earn more income, nurture their families, achieve personal growth, or live more spiritual lives. And, though the survey doesn't reveal whether some aspirations might be expressed in dubious ways, no one aspires to crime and corruption for its own sake.[4] Aspirations urge us forward in the epic human quest against complacency.

Second, aspirations are intrinsic, even if they are influenced by external factors. Agyare wasn't shackled by the golden handcuffs of the tech industry. She left when she could no longer ignore the beat of her inner drummer. Psychologists Edward Deci and Richard Ryan defined the related concept of intrinsic motivation as "the prototypical form of self-determination: with a full sense of choice, with the experience of doing what one wants, and without the feeling of coercion or compulsion, one spontaneously engages in an activity that interests one."[5] It's not an aspiration if a person has to be coerced or asked. Aspirations come from within.

Third, aspirations are slow and sticky. They sustain for the long haul. Intrinsic growth doesn't happen in a day, so any force that doesn't last won't be enough to inspire it. Agyare's path from student to entrepreneur and activist took eleven years, during which she expended consistent effort. Something had to pull her through the demanding course load at Ashesi, and something had to keep her going at work–not just to earn income, but to learn and grow. Of course, throughout that time, other internal forces–needs, fears, desires, irrational impulses; reasoned goals to be richer, kinder, healthier, and happier–competed for her attention. But these were fleeting compared with the pull of her primary aspirations. Psychologist Kennon Sheldon has spent much of his career researching the value of setting and striving toward intrinsically motivated, or self-concordant, goals. He and his colleagues have found that "individuals do better at self-concordant goals because they put more sustained effort into such goals."[6] Take the long view and average out the choppy waves of our minute-by-minute moods, and what remains are the slow tidal swells of aspiration.

Fourth, it was by chasing her aspirations that Agyare underwent critical intrinsic growth. She admits that even if she had wanted to start Soronko earlier in her life, she wouldn't have been ready. It's not an easy thing to start your own company. And to succeed at both a for-profit and a nonprofit at once is another thing altogether. What equipped her with the skills and strength she needed?

What she learned through her early aspirations enabled her later ones to develop. Agyare's first aspiration helped her gain knowledge and professional discernment. Building a new organization requires business acumen, management capability, and leadership skills, all of which Agyare learned in her first career. It also seems likely that years of academic life and professional work further built up her self-control. Ashesi routinely graduates high achievers. Corporate high-tech Ghana is much like the high-tech world elsewhere: It values a hard-charging work ethic. That mental and emotional stamina supported Agyare as she built Soronko. Following her aspirations led to a sharper mind and greater will.[7]

It's important, though, that in meeting one's aspirations, some striving is involved. An aspiration achieved without effort doesn't build wisdom. "Undeserved" fame or fortune doesn't necessarily cause growth, because they're not accompanied by internal change. The spoiled children of inherited wealth are not particularly wise. The same problem occurs at a national scale when the resource curse of oil and minerals corrupts leaders and stunts other industries.[8] Even more stable countries are prone to "Dutch disease," where the availability of an easy resource displaces other productive capacity, just as an overused crutch can lead to muscular atrophy.[9] Apparent exceptions only affirm the rule. There are trust-fund children who increase the prestige of their families, but they're focused on more than collecting baubles and living a lavish social life. Among nations, there is, for example, Norway, which took a windfall from North Sea oil, invested it carefully, and pumped some of the returns into one of the world's most generous foreign aid programs.[10]

That striving after aspirations leads to greater mind and will is not surprising. Intentional effort leads to learning. What's most noteworthy about Agyare's intrinsic growth is her change in heart. She underwent a fundamental change in intention–from being focused primarily on serving her own economic, intellectual, and emotional needs through a corporate job to becoming increasingly focused on serving the needs of others. "It took me four years and a lot of sleepless nights to realize that when you have a dream and a desire, it is like an alarm clock going off inside you," she said. "Hitting the snooze button doesn't work; the alarm will just go off again. Eventually, you have to wake up."[11]

A New Beginning

Long before Agyare started at Ashesi University–years in fact before there was an Ashesi at all–Patrick Awuah vowed that he would never return to live in Ghana. He was visiting his home country in 1990 when he became disgusted by its political corruption, economic stagnation, and social backwardness. Five years earlier, he was among the lucky few who, due to a combination of wise parenting and strong academic accomplishments, earned a full scholarship to Swarthmore University in the United States. His dream then was to become an engineer and to buy a new house for his mother. He majored in electrical engineering and economics. Upon graduation he joined Microsoft as a program manager. Over ten years, he proceeded up the corporate ladder rung by rung.[12]

That decade marked the golden years of Microsoft. It was a period when Awuah and his colleagues worked day and night to realize the company's mission to put a PC on every desk. The market responded with an industry boom whose reverberations grow louder even today. Through stock and options, the company shared its skyrocketing profits with its employees. The first Microsoft millionaires were minted.

By 1995, Awuah was comfortably among them. Having achieved and exceeded his dreams of just a decade earlier, though, he began to ask if there wasn't more to life than technical feats and his family's financial well-being. He felt incredibly blessed and wanted to share his good fortune. Reflecting on his life, he felt a critical point for him was his Swarthmore education. Its strong liberal arts program marked the point where his path and that of his high-school peers diverged. Based on this conviction, and with encouragement from his wife, Awuah decided to start a liberal arts college in Ghana. In 1998, he quit Microsoft.

He enrolled in UC Berkeley's Haas School of Business, where he focused every class project on the question of how to start the university. In 2002, spurred by a quotation attributed to Goethe–"Whatever you can do or dream you can, begin it. Boldness has genius, power and magic in it!"[13]–Awuah launched Ashesi University in Ghana's capital city of Accra. Defying his earlier vow, he moved back to the country with his family soon afterward.

"Ashesi" means "beginning" in Fanti, the language of Awuah's ancestors, and the name has proven auspicious. Starting with the initial class of 25 students, whom I taught in rented space, the school has grown to 600 students studying in a beautiful new campus. Talented students flock to Ashesi not only from within Ghana but from across West Africa. International development experts cite the school as an example of what can be accomplished by a country's own people.[14] Other private universities have sprung up in Africa to follow its lead.

Ashesi is an unqualified story of successful social change. But while Awuah's professional life is steeped in computing technology, the lessons of his life don't have much to do with the tools he used. What matters is his transformation from a young man eager to study abroad, to a program manager in a large corporation, to a visionary who invested his wealth, sweat, and soul in a cause that is changing Africa. In other words, it's not about the packaged intervention.

Transformational Internal Epiphanies

Something happens as aspirations are achieved through valiant effort. When dreams come true, some people notice that they've outgrown them. Awuah says that in his last year at Microsoft, "I lost the sense of urgency that I had before. The work just felt less important." People talk of missing something in their lives and of wanting something more, something they might have postponed or never have imagined before. They feel an intrinsic change. It's not that they gain new knowledge. Don't we all know, after all, that there's more to life than whatever we happen to be chasing right now? Rather, they move on to a different, deeply felt aspiration. At some point, Awuah was no longer attracted to the extra 10 percent of income, recognition, or accomplishment that he had spent years pursuing. He wanted work with broader purpose.

These kinds of stories are common fodder for heartwarming news articles, but they are rarely discussed by policymakers. No one in the US government, the World Bank, or the Gates Foundation is asking, "How do we encourage people through transformational internal epiphanies?"

Partly, the problem is that today's metric-focused technocrats all but laugh at what seem like soft intangibles. Partly, the problem is that policy is disconnected from fields that consider these changes of heart. Partly, it's that the fields that once used to think about deep questions like this have stopped asking them. So much of behavioral science today overlooks long-term human change in favor of easily measured short-term phenomena. It focuses on catchy factoids, such as that finding a dime left in a phone booth can make you temporarily more likely to help others.[15] As a result, modern social policy is obsessed with the equivalent of strategically placed coins–tricks and nudges to incentivize "behavior change."[16] But while behavior change might be a more meaningful goal than the willy-nilly scattering of technologies, it's fleeting. And it casts individuals as adversaries to be manipulated, as if people just can't be trusted to do the right thing on their own. The alternative is to ask, What makes a person intrinsically motivated for the larger social good? What makes a Patrick Awuah or a Regina Agyare? Today's number-crunching disciplines have no answer to such questions.

But developmental psychology does. Psychologists going back to Sigmund Freud have sought to explain human maturation as a staged process of *personality*, *character*, or *life-cycle development*.[17] Freud was joined by many others. Thinkers such as Erik Erikson, Jean Piaget, and Lawrence Kohlberg tried to map out aspects of long-term human development.[18] Some of their claims have been discredited by modern psychology, but in acknowledging the possibility of life-long, intrinsically powered growth, they offer an alternative to today's fast-twitch policymaking.

And they are far better suited to explaining transformations such as Awuah's. His trajectory was defined by his successive aspirations. As a student he "wanted to be a great engineer. My father was a mechanical engineer, and I'd read magazine articles about things like the Space

Shuttle. I thought it was cool." At Microsoft he did exactly the kind work he once admired. "Growing up in Ghana, whenever I read about technological advances, they were always happening abroad," he said. "In my job, though, I was at the center of it. We worked on computer networking when it was going mainstream. I knew some kid would read about what I was working on and think, 'That's awesome!'" Awuah says he came to love "the technical challenges of the work and the intensity of the workplace."

Then, after nearly ten years of that work, "being in a meeting to decide what a button should do or what feature to cut stopped seeming that important." What happened? We've already encountered Agyare, who after years of corporate work, struck out on her own. Similarly, Rikin Gandhi left a software engineering job at Oracle and astronaut dreams to found Digital Green. Abraham George returned to India to establish Shanti Bhavan after attaining entrepreneurial success in America. Trish Millines Dziko was a Microsoft employee before she went on to start the Technology Access Foundation. Bill Gates made a public transition from amassing one of the world's largest fortunes to spending it on global philanthropy. And when I ran the research group in Bangalore, I received inquiries every week from professionals who wanted to work with us. They would say, "I've done well for myself, but I'd now like to see how I can give back to society." There's a good chance that you know people like this, or maybe you are one.

Without taking away from anyone's uniqueness–each of these people is a gem on the pebble beach of humanity–there's no doubt that a pattern is at work.[19] A pattern of human maturation. A pattern of intrinsic growth. A pattern of expanding heart, mind, and will.

Maslovian Development

Developmental psychology has many theories that could explain this pattern. The one that best fits what I've witnessed is the well-known hierarchy of needs developed by the psychologist Abraham Maslow. Lists of the world's eminent psychologists inevitably include Maslow, and his ideas have taken firm root in fields well beyond psychology.[20] But if

Maslow's hierarchy is a bit of a celebrity meme, it has suffered for its popularity–it's widely misunderstood. Most people have heard about the hierarchy from pop psychology, and while some of what they've heard is correct, it's mixed with inaccurate rumor-mongering. For example, most depictions of it involve a rainbow-colored pyramid that Maslow never used. The wrong interpretations don't explain Awuah or Agyare.

In its original form, Maslow's hierarchy was a series of five motivational categories: Survival needs such as hunger, thirst, and the need for sleep motivate a person to seek food, water, or shelter. Security needs drive a person toward physical and psychological security–freedom from fear and anxiety, desire for structure and order, and so on. Needs for love and belonging are met through social acceptance, community, and companionship. Esteem needs demand recognition, status, achievement, and competence. And the need for self-actualization causes people to do things that express their unique talents and preferences, what "he or she, individually, is fitted for," whether it's playing in a rock band, managing a corporate division, or being a good parent.[21]

Maslow suggested that these needs–which everyone shares–were sorted by their urgency: "It is quite true the humans live by bread alone–when there is no bread. But what happens to their desires when there *is* plenty of bread and when their bellies are chronically filled? *At once other (and higher) needs emerge* and these, rather than physiological hungers, dominate the organism. And when these in turn are satisfied, again new (and still higher) needs emerge, and so on" (emphasis in original).[22]

As immediate hunger is satisfied, more security is sought; as security needs are satisfied, esteem becomes more important; as the need for esteem is satisfied, self-actualization becomes a stronger motivator.

Less understood is Maslow's claim that people are "multimotivated" and that behavior is "multidetermined." People have multiple needs at once, even if one might dominate. At any time, each of a person's needs might be satisfied to different degrees: "It is as if the average citizen is satisfied perhaps 85 percent in physiological needs, 70 percent in safety needs, 50 percent in love needs, 40 percent in self-esteem needs, and

10 percent in self-actualization needs."[23] And any single behavior might be motivated by multiple needs. We work hard at our jobs because pay and benefits satisfy physiological needs, security needs, and esteem needs; the recognition for our work satisfies esteem needs; and, if the work is deeply interesting to us, it is an expression of self-actualization. Depending on the nature of the job, the reward package, and one's attitudes, each of these components will have different weight.

Maslow Revisited

Maslow has had his critics, and they have questioned everything from the number and substance of his levels to their ordering, their individualistic focus, and their bias toward his own gender and culture.[24] (Personally, I'm not convinced that Maslow's "belonging needs" are a single level in the hierarchy as much as a set of needs that runs parallel to the others.) Some of these criticisms are valid, but many are based on a poor reading of Maslow.[25] A truer understanding explains how individual intrinsic growth such as Awuah's can happen.

One pervasive mistake is to see self-actualization as the top of the hierarchy.[26] Maslow kept reworking the hierarchy throughout his life. He wondered, for example, whether esteem needs might be further broken down into two levels–one for public recognition and another for private achievement or mastery.[27] And more relevant for Awuah, Maslow also realized that "self-actualization is not enough." He gestured at an additional level that could be called *self-transcendence*.[28] Self-transcendence tends toward "the good of other people," toward egolessness and altruism.

Self-transcendence is essential to explaining Awuah's evolution, because he didn't stop with self-actualizing work. Awuah says that toward the latter half of his time at Microsoft, he felt a growing desire to contribute to the larger world. "The birth of my son as well as events in Africa" catalyzed his transition, he said, referring to the Rwandan genocide and the Somalian civil war. His life path shows the clear pull of self-transcendence overtaking needs for security, esteem, achievement, and self-actualization.

Yet another misreading of Maslow's theory goes like this: Some people go on hunger strikes to protest injustice; their physiological needs don't prevail over other motivations; so something must be wrong with the hierarchy.[29] Actually, though, this example offers proof of something else. Maslow knew that there were both intrinsic and extrinsic causes of behavior and said that different people felt the pull of lower needs differently: "It is precisely those individuals in whom a certain need has always been satisfied who are best equipped to tolerate deprivation of that need in the future."[30]

So Maslow's hierarchy is actually two hierarchies. One–about the influence of external conditions on behavior–explains why most of us would prefer to go three days without self-actualizing work than three days without water. That's the hierarchy of needs as it's popularly understood. The other hierarchy–about intrinsic motivation–explains what allows some people to sacrifice lower needs for the sake of higher ones. A hunger striker puts self-transcendent goals ahead of survival needs. That shows a kind of maturity, an ability to suppress or ignore more urgent needs in the service of a larger aspiration. To distinguish this internal gradient from the external hierarchy of needs, I'll refer to it as the *hierarchy of aspirations*.[31]

Maslow saw this internal growth as a good thing. He called it "character learning" or "intrinsic learning," which is part and parcel of intrinsic growth.[32] Improvements in intention, discernment, and self-control allow a person to act not just in pursuit of pressing, self-focused, short-term needs, but also toward longer-term outcomes that may enhance others' well-being. At the lowest level are physiological aspirations, which are narrowly self-, present-, and material-focused. Someone operating primarily at this level sees the world as a raw fight for survival. That narrow focus is stretched with security aspirations, however, which are still self-focused but require some future planning and cooperation to obtain. Next come esteem and achievement aspirations, which begin to incorporate nonmaterial needs and further instill longer-term focus and the drive to cooperate.[33] Still, these aspirations remain basically selfish, and selfishness continues all the way through self-actualization. In fact, self-actualization could be thought of as the height of selfishness. Self-actualizers are most interested in ensuring their own long-term happiness–materially, intellectually, and emotionally. (This might be one reason why various studies suggest that today's creative classes seem more narcissistic than ever.[34]) But their selfishness is not like the selfishness of brutish survival–it can express itself as enlightened self-interest. Aspirants to self-actualization are often vocal in protecting universal freedoms, because they have both the slack and the desire to guarantee their own right to self-expression. Lastly, there are self-transcendent aspirations whose focus on self gives way to genuinely selfless concern for others. All the way through, intentions embrace larger and larger circles of humanity. With greater Maslovian development, people become more future-oriented, more other-oriented, and more other-oriented toward larger groups of people.

Awuah is a good example. He became less interested in esteem and self-actualization once he could take them for granted. Something internal changed as he learned new skills and grew more confident in his ability to excel in a satisfying career. Maslow suggested that these shifts happen when a need "has always been satisfied." A more robust explanation, though, would be that shifts occur when a person feels thoroughly confident in being able to satisfy a need, either due to his or her own ability or through societal provision.[35] In any case, it's because Awuah is motivated mainly–though probably not solely–by higher needs that lower needs cease to exert strong pull. He has skipped many lunches for the sake of establishing a new university. He not only let go of a high-income career, but also spent much of his wealth on the cause. And, most telling, whatever sacrifice he felt–"I won't lie to you–it was tough" to leave Microsoft, Awuah said–it didn't get in the way of his new aspiration.[36]

In contrast, someone who aspires primarily to survival or security would not feel comfortable making Awuah's tradeoffs. A shocking illustration is a story told to Kevin Bales, an expert on global slavery and president of the nonprofit Free the Slaves. An Indian couple who were locked in bonded labor came into an inheritance that allowed them to pay back a familial debt and buy their freedom. But then, the husband says: "We paid off our debt and were free to do whatever we wanted.

But I was worried all the time–what if one of the children got sick? What if our crop failed? What if the government wanted some money? Since we no longer belonged to the landlord, we didn't get food every day as before. Finally, I went to the landlord and asked him to take me back. I didn't have to borrow any money, but he agreed to let me be his [bonded plowman] again."[37]

This couple had freedom, but then voluntarily returned to conditions of slavery for the sake of guaranteed food and security, proving the powerful pull of Maslow's physiological and security needs when both are under systemic threat.[38] Aspirations for achievement or esteem, let alone self-actualization and self-transcendence, are nowhere in sight. Some people seek greater income at the expense of a satisfying career; others don't mind low pay if the job fulfills a deep creative urge. Some people are magnanimous only when they're acknowledged for it; others are consistently generous, indifferent to recognition. These kinds of differences are explained by the hierarchy of aspirations.

One thing about the hierarchy is that it allows for the twisty evolution of aspirations. Take the example of some poor communities that pressure their members to share material goods. They make it difficult for individuals to accumulate wealth. This makes sense for collective survival in severe circumstances, but it can discourage private effort. A social norm that instead respects personal property rights can motivate individuals, and it encourages material growth. At some level of prosperity, though, selfish accumulation can lead to stark inequalities that strain the social fabric. Sharing once again becomes important, though perhaps in a less personal or communal way. Thus, intrinsic growth can be a climb with switchbacks: Progress means that sharing will give way to private ownership, which evolves into enlightened sharing. Similarly, people grow from dependence to independence to interdependence; from unwanted poverty to prosperity to contentment; from oppression to freedom to responsibility; from helplessness to confidence to humility. The qualitative changes inherent in the hierarchy allow a Hegelian back-and-forth that constantly strives for balance, synthesis, and maturation.

The climbing analogy helps clarify other points. First, the hierarchy doesn't imply inexorable upward motion. It's possible to regress, stagnate, or vacillate. At most, the hierarchy is a map, and maps by themselves don't decide where people go. Also, the map says little about the flora and fauna you might encounter. The hierarchy of aspirations only captures one aspect of human personality relevant for social causes. The theory leaves everything else about our infinitely rich behavior unexplained and unconstrained.

Individuals also begin at different altitudes and climb at different speeds. Awuah was raised by parents who weren't struggling for survival, so he took subsistence for granted–undeniably a more advantaged starting point than, say, Isaac Tuggun's. Other people start with self-actualization as a given. They might proceed right into self-transcendence. This suggests that children raised to be comfortable with one set of aspirations can aspire for more. In a letter to his wife, John Adams, America's second president, wrote: "I must study Politicks and War that my sons may have liberty to study Mathematicks and Philosophy. My sons ought to study Mathematicks and Philosophy, Geography, natural History, Naval Architecture, navigation, Commerce and Agriculture, in order to give their Children a right to study Painting, Poetry, Musick, Architecture, Statuary, Tapestry and Porcelaine."[39] The progression goes from the hard and practical to the constructive and exploratory and eventually to the artistic and self-expressive.[40] More than a century and a half before Maslow, Adams mapped out a path for his offspring from security to achievement, and from achievement to self-actualization.

Confronting the Tech Commandments

As I've worked with poor and marginalized communities, I've encountered a wide range of amazing people: alcoholic men who became virtual teetotalers after taking up meaningful work; an ostracized victim of gang rape who started an organization that rescued prostitutes; former ethnic antagonists who came together to build a hospital; an engineering graduate who moved to an impoverished village he supported for decades; and more than a few destitute nonprofit beneficiaries who made a transition to capable and reasonably paid nonprofit staff. These personal metamorphoses aren't frequent by any means, but neither are they altogether rare. When they occur, they signal an evolution like Agyare's and Awuah's. They're accompanied by a surge of intrinsic growth. They inspire others. And they shine and shimmer against the inertness of packaged interventions.

What's missing in today's main paradigms of social change is any notion of intrinsic human progress. You might admire Tuggun and Sreenivasa for their ability to lead independent lives despite hard initial conditions. Or you might praise Agyare and Awuah for their larger contributions to society. Either way, what you're really saying is that the world needs more people to become better versions of themselves. But we have no basis for acting on that idea without a framework of internal human betterment.

The dominant voices in public policy model people as having the same fixed preferences.[41] Most economists, for example, think about how to tune external incentives to produce mass changes in behavior.[42] Market mechanisms are exalted precisely because they sculpt the supposedly granite hardness of human greed into architecture that uplifts us all. Few think about causing long-term changes in society through growth in individual character.[43]

But even economists' most vocal critics–qualitative scholars who emphasize culture, context, and complexity–don't offer frameworks of individual progress. Some cultural anthropologists instead celebrate the diversity of human behavior that results from a diversity of context. Under their cultural relativism, human beings adapt intelligently to their geography, history, culture, and structures of power. All peoples are equally worthy, and there is little allowance for either personal or societal progress.[44]

The economic and anthropological models of human behavior sit at opposite extremes of the social science spectrum–the fields are often at odds in the way of snakes and mongooses–but they still have one trait in common: They assume that behavior depends largely on external context, so they focus on changing external circumstances while neglecting intrinsic growth.[45]

The same neglect happens in politics. During America's 2012 presidential debates, the candidates presented what they felt were their most appealing positions. Obama captured the left's emphasis on external forces outside of the person: "The federal government has the capacity to help open up opportunity and create ladders of opportunity and to create frameworks where the American people can succeed." Representing the right, Mitt Romney said, "The primary responsibility for education of course is at the state and local level. . . . Every school district, every state should make that decision on their own." The left is afraid of blaming victims and downplays individual intrinsic growth. The right leaves communities short on intrinsic growth to fend for themselves.[46] Either way, intrinsic growth is neglected where it's most needed.

That neglect lines right up with the Tech Commandments, which demand that we twiddle external circumstances and steer clear of wrestling with values. But social inequities cannot be eliminated with value-free changes alone. Disparities are due in part to individual differences in traits to which we ascribe value. It cannot be true both that intrinsic growth matters to a good life *and* that more intrinsic growth isn't somehow better. We can debate what makes a person good, and we should. But pretending that everyone is equally good, or dismissing virtue-building policies as those of a "nanny state," undermines attempts to foster heart, mind, and will.

Imagine if, with the wave of a magic wand, the wealth of everyone on the planet were suddenly to increase, and imagine the windfall happened progressively, so that dollar-a-day people had their wealth multiplied by ten, while billionaires saw only a 1 percent increase. With that one spell, you'd have generated economic growth, more equality, greater dignity, increased freedom, and instantaneous happiness. Yet it is not at all clear that the world will have become a better place in the long run. The happiness would fade, we'd consume even more, and poorer people would slip back into poverty. Suppose, instead, that the spell created a sudden increase in everyone's heart, mind, and will, with no other change. That would tend to lead naturally to economic growth where desired (and less where not), more justice, greater dignity, more freedom with responsibility, and increased likelihood for enduring global well-being.

Intrinsic growth stresses internal maturation over external change. It accepts that real progress is slow and gradual. And it links progress to the flourishing of certain universal values–not only of personal freedoms, but of personal goodness. A framework of human development provides a counter to the Tech Commandments. Real progress isn't strictly about satisfying our every present desire. It's about our desires themselves evolving.

CHAPTER 9

"Gross National Wisdom"

Societal Development and Mass Intrinsic Growth

Progress can be a dangerous idea. It can mean labeling people with scores, and people read a lot into them. Labels discourage or insult low scorers, induce arrogance and complacency in high scorers, and cause society as a whole to warp with prejudices, none of which is helpful.[1] What's worse, the claim that progress is dependent on personal qualities such as heart, mind, and will begins to sound a lot like blaming the victims of poverty, oppression, and prejudice, as if lack of intrinsic growth were their fault.

This is an incendiary issue, and I believe we have put off for too long an intelligent conversation about it. One way to tease out the tough ethical issues is through the story of a high-school student I once tutored. David was born into a poor, dysfunctional family and became the foster child of wealthy parents. They provided a stable home and enrolled him in an expensive private school, where he was respected as a star athlete. But he was behind in all of his classes. Working with him on geometry, I found he was able to do the math if he put his mind to it, but, without adult prodding, he wouldn't. He didn't have much motivation to study (intention), didn't seem to think it worthwhile (discernment), and, in any case, didn't apply himself to his assignments (self-control).

I don't blame David for his poor showing in geometry, or even for his lackluster heart, mind, and will. Nor should anyone else. His birth family might have been cursed with drug abuse, alcoholism, homelessness, illiteracy, insolvency, emotional turmoil, or just garden-variety bad luck or bad judgment. You can't blame a child for not developing good study habits under constant distress. Yet Walter Mischel's famous "marshmallow study" showed that the capacity to delay gratification–a kind of self-control–expressed at ages four to six is among the strongest predictors of achievement and social adjustment in young adults.[2] It doesn't make much sense to hold six-year-olds accountable for their personalities–it's obviously not up to them. But what Mischel's research further implies is that the responsibility for an adult's degree of self-control isn't black or white, either.[3] A person's intrinsic growth is never wholly of his own making.

But, even if David wasn't to blame for deficiencies of heart, mind, and will, he needed more intrinsic growth to be better prepared for a better life.

Meanwhile, we need more intrinsic growth from richer and more powerful people as well. Accustomed as they (we?) are to comfortable cocoons, they maintain self-absorbed, self-satisfied lifestyles without expanding their circles of intention. Recent studies suggest that, compared to those with lower social status and less income, people with greater status or income express less empathy, act less ethically, and give proportionately less to others in need.[4] Whatever intrinsic growth influential people have–and they don't deserve all the credit for it any more than David deserves blame–more is needed. Due to the power they wield, their further intrinsic growth is essential for societal well-being.

If individuals are reluctant or unable to cause their own intrinsic growth, though, what could encourage them? I've emphasized individual heart, mind, and will so far, but nothing in this book is meant to suggest that our lives are entirely within our own personal control.[5] External conditions matter. Social structures matter. Agyare and Awuah both credit their initiative to supportive parents who always encouraged their endeavors. Sreenivasa's life would be very different if she had been born not in rural South India, but in Sweden, Soviet Russia, or Saudi Arabia.

But external circumstances as an individual perceives them are often under human control as society's collective actions. The legal freedoms you enjoy, the tax rate you pay, the public education available to you and your neighbors, and the culture in which you are immersed are not products of decisions you make on your own, but they are human choices nonetheless.[6] Would individuals be better off with improved health care? Better education? Stronger communities? Would citizens be better off with wiser leaders? Future-focused policies? Separation of wealth and state? Absolutely. But these require better intention, discernment, and self-control at the level of whole societies–what you might call *societal intrinsic development.*

Of course, societies are infinitely complex, and social norms are notoriously difficult to alter. Large-scale social problems, though, are not intractable. Change may be slow, hard, and entangled in complex issues, but it happens. Most of us have witnessed it within our lifetimes, which should give us hope that mass social change is not a fluke. By considering cases of societal intrinsic development, we can see how they're connected to individual intrinsic growth.

A Billion Shifts in Aspiration

William Blake saw the world in a grain of sand; I saw India in a taxi driver. I met Narasimha through a dial-up cab service one day when I needed to go to the airport. I was used to drivers who called five minutes late only to tell me it would be another ten minutes. So this time I had specified a much earlier pickup than I needed. Narasimha, however, showed up ten minutes ahead of time. On the road, he was unlike other Bangalore drivers. He didn't seem to have an insistent death wish. He was safe and steady.

We started a conversation, but when we hit the limits of my Kannada, Narasimha switched to Hindi. When that, too, proved insufficient, he switched to broken English. I learned that he was unmarried and came from a village a couple of hours outside of Bangalore. His family farmed, but an uncle had introduced him to the taxi business. Narasimha's dream was to own his own car, and he had nearly saved what he needed to purchase one. As we approached the airport terminal, he handed me a business card and asked me to call him whenever I needed a ride. By calling him directly, I would be cutting out the middleman.

And so I did. Narasimha worked smart and hard. I sometimes had flights leaving at two o'clock or six o'clock in the morning, but he never turned down the 30-kilometer ride to the airport. Sometimes he was committed to multiday tours outside of Bangalore and would call up one of his "brothers"–friends from his network of fellow drivers–to substitute. He would then call me before the meeting time to let me know the driver was coming and confirm during the ride that I had been picked up. His friends all got me to the airport, but none as promptly, safely, or reliably as Narasimha.

One day Narasimha arrived in a shiny white Ford Icon. It was the first car he owned himself, and it meant that he was his own man. He no longer had to rent cars from the cab fleet. I happened to be his first ride, and he beamed when I congratulated him. He pampered his vehicle, and on each subsequent trip, I noticed he had acquired something new for it: a carpeted cover for his dashboard, rear speakers, blingy LEDs for the deity on his dashboard, and eventually a stereo with a CD player. I asked Narasimha, now that he had his own car, what he wanted most in his life. He said he wanted to trade up to a larger vehicle. He also wanted to find a suitable husband for his younger sister–he had to raise enough dowry to be able to do this. It seemed the next generation of Narasimha's family would be raised in a world very different from the one in which he had been brought up.

Narasimha is part of a national shift in India. He represents a new stratum of Indian society that is rapidly climbing out of poverty. It's driven in part by the country's growing wealth and in part by a wave of mass intrinsic growth. Although Narasimha grew up in a poor rural family, the combination of basic education, self-motivation, and economic opportunity puts him in a new category. He's somewhere between the vast population of the extreme poor who haven't yet seen much change and the historical upper classes. He and his peers are transitioning en masse from low-income survival to middle-class security.

In the sociological classic *The Protestant Ethic and the Spirit of Capitalism*, Max Weber suggested that Christian Protestantism, and specifically its tendency to see worldly work as virtue, provided the spark for modern capitalism. Had he witnessed what I saw in Narasimha, though, he might have reconsidered his thesis. Ironically, the "Hindu rate of growth" was once a term of derision used by Indian intellectuals to complain about the country's 1.3 percent per year growth in per capita income, a snail's pace that lasted from independence in 1947 through the 1980s. In the 1990s, though, the Indian economy took off. From 1992 to 2010, India's GDP growth averaged 6 to 7 percent a year, at times hitting 8 or 9 percent even during the global recession. If Weber were born a hundred years later and 6,000 kilometers southeast, he might have overlooked Protestant worldly asceticism as the cause of rational industry, and instead pointed to a Hindu ethic to work uncomplainingly at the task at hand. Or, if he had visited modern China, he might have linked Confucian discipline to modernization.[7]

With hindsight, it seems clear that the cause of economic growth is not Protestantism or Hinduism or Confucianism as such. There's something in the human organism that compels us to aspire beyond survival and security when conditions are right. People have intrinsic aspirations for productivity and self-expression.

Financial incentives undoubtedly play a role, but, as Weber was careful to point out, aspirations are not strictly about economic gain. Weber felt that there was something else: "It is an obligation towards the content of his professional activity, no matter in what it consists."[8] This earnestness infused my taxi driver. Narasimha seemed particularly eager to please his customers. If I had any luggage, he would rush to relieve me of it. He was always sure to ask whether I wanted air conditioning when I got in his car. And if I asked for music, he was ready with a rack of hand-labeled CDs ranging from South Indian Carnatic music and Bollywood hits to American rock-and-roll and European techno-pop. If more than a few weeks went by between trips, he would call me to ask whether he could be of service.

In 2009 I told him that I was moving back to the United States and that it might be a while before I would be back. When we arrived at the airport, I tried to pay him the fare as well as a bonus for his years of reliable service, but he initially refused both. With tears in his eyes, he said he just wanted me to call him the next time I was in town. I finally had to stuff an envelope of cash in his shirt pocket and run off promising that I would. A similar scene repeated itself six months later when I returned to Bangalore for a brief trip. Narasimha again refused payment until I insisted that it was important for his business. Weber was right: There was a sense of purpose in Narasimha's hard work that went beyond profit.

Multiplied by the billion-plus population of India, these changes are manifested countrywide as budding entrepreneurship, increasing consumption, expanding political empowerment, and heightening national pride. Middle-class India is undergoing rapid change. Each year, fewer young people are interested in a stable job at a large corporation, and want instead to make their name through a start-up. Political activism has crescendoed, with a younger generation eager to uproot government corruption. Meanwhile, hordes of people now throng to India's malls. When I arrived in Bangalore in 2004, there were two malls in the city; when I left five years later, there were more than ten, all bursting at the seams with a brand new generation of shoppers. And on September 24, 2014, the Indian Space Research Organisation announced that it had put a spacecraft, the *Mangalyaan*, into orbit around Mars. National newspapers quoted the prime minister proclaiming, "This is [the] first time in the world. History has been created today."[9]

Evolving Mass Values

India's changes recapitulate elements of developed-world history. Like India today, the United States was buoyed by an entrepreneurial spirit. A battle against America's rent-seeking political corruption was fought effectively in the nineteenth century.[10] Consumer culture, now burgeoning in India, was all but invented in the United States circa 1920. And Indian pride in their Mars orbiter echoes American sentiment toward Apollo 11, the world's first moon landing.

Despite huge differences in culture and history, there is no denying the commonalities among societies that experience socioeconomic growth. I grew up partly in Japan and partly in America. Despite the differences between the two countries, the repeating patterns of modernization were evident: Infrastructure tends to improve, technology to proliferate, government to become more efficient, parents to have fewer children, industrial employment to overtake agriculture, and gender disparities to shrink.[11]

When I landed in India, I felt a sense of déjà vu. There was something of the mid-twentieth-century Japan that my parents and grandparents had spoken of. City streets bustled with an energetic commotion. Individual aspirations–like Narasimha's–were woven tightly into the country's development. Ads for financial services urged people to "Invest in your family, invest in India!" There was a sense of national mission.

To believe in progress doesn't mean that any country has arrived at the final destination, or even that there is agreement about a single destination. For example, any claim that contemporary Western society is the apex of human civilization is unwarranted–America, to take one case, could surely do better than a 13 percent rate of poverty, a government captured by moneyed interests, a stubborn resistance to addressing climate change, and reality TV. But just because progress is hard or variously defined shouldn't mean we should scrap all hope of it.

Consider the change in women's status in so many countries in just the past half-century. In 1977, two-thirds of Americans believed that women should not work outside of the home; by 2012, less than one-third of Americans thought so.[12] In the 1970s, women's median full-time earnings were 60 percent of men's; by 2011, they had risen to 77 percent.[13] In 1970, women held 11 of 535 seats in Congress–just 2 percent; in 2014 women took 100 seats for the first time, nearly 20 percent.[14] We still have a long way to go, but advances have been steady, self-sustaining, and large-scale.

To even the casual observer, it should be clear that these kinds of transformations didn't happen through packaged interventions. Some argue that household appliances and contraceptive technologies revolutionized the role of women in society. But these inventions, as with Amazon's effect on the book industry, were accelerators and amplifiers, not primary causes.[15] The pill was approved for use in America in 1960, but US women's movements go back to at least the mid-1800s. Women's suffrage was achieved in 1920. The Equal Pay Act was passed in 1963. And all through these milestones was the constant fight for equality, fought on the frontlines by women themselves and supported by national shifts in attitudes among both sexes–all forces that are difficult to measure and difficult to package.

Other countries have followed a similar course. In the late 1960s, my mother was 1 of only 5 women at Keio University's School of Medicine in a class of 95, so women made up only about 5 percent of the total. Today, there are about 90 women in each class of about 450, making it 20 percent. That's progress–even if, again, there is more to do. The United Nations Development Programme finds that, between 2000 and 2013, 96 percent of countries tracked saw reductions in overall gender disparities based on indicators of health, education, political representation, and employment.[16]

So, it seems that individual intrinsic growth piles up and manifests as large-scale societal development. But is there hard evidence?

Remarkable data comes from Ronald Inglehart, a political scientist at the University of Michigan.[17] Inglehart is the founder and former president of the World Values Survey, a series of large-scale surveys of national values that he and his colleagues had the foresight to begin in the early 1980s. It spans four decades, all six inhabited continents, 97 countries, and 400,000 respondents. The survey asks about people's values, beliefs, and aspirations. It considers work, family, religion, happiness, government, and environment.[18]

Over the years, Inglehart and his colleagues have dug deep into this mountain of data and found striking patterns. They found, for example, that mass individual aspirations correlate with modernization, economic development, and democratization. Here is the heart of their conclusions, excerpted from the World Values Survey website:

- Much of the variation in human values between societies boils down to two broad dimensions: a first dimension of "traditional vs. secular-rational values" and a second dimension of "survival vs. self-expression values."

- Traditional values emphasize religiosity, national pride, respect for authority, obedience, and marriage. Secular-rational values emphasize the opposite on each of these accounts.

- Survival values involve a priority of security over liberty, non-acceptance of homosexuality, abstinence from political action, distrust in outsiders, and a weak sense of happiness. Self-expression values imply the opposite on all these accounts.

- People's priorities shift from traditional to secular-rational values as their sense of existential security increases. The largest increase in existential security occurs with the transition from agrarian to industrial societies. Consequently, the largest shift from traditional toward secular-rational values happens in this phase.

- People's priorities shift from survival to self-expression values as their sense of individual agency increases. The largest increase in individual agency occurs with the transition from industrial to knowledge societies. Consequently, the largest shift from survival to self-expression values happens in this phase.[19]

In other words, changes in individual aspirations are connected to the visible aspects of national modernization. These transitions have obvious correlations with Maslow's hierarchy. Traditional values would be expected of people in subsistence agriculture or servitude, who might habitually live on the edge of unfulfilled physiological needs. If caprices of nature or human authority sway outcomes far more than individual effort, it seems only natural that people would fail to develop any sense of their own ability to control outcomes.

The slack generated from greater agricultural productivity, though, sets the stage for industrialization and the spread of secular-rational values.[20] Factories and other forms of formal work reinforce secular-rational values by providing day-to-day regularity and the beginnings of a meritocracy. Industrialization, however dehumanizing, is a reprieve from the much harsher whip of luck, weather, and subsistence living.[21] This transition from traditional to secular-rational happens while survival values and security aspirations remain strong, but it also leads to an appreciation of goal-directed hard work, achievement, and esteem.

When stable jobs have secured material well-being, people seek more autonomy, independence, and self-expression. These trends grow a larger class whose main aim is self-actualization. Inglehart and his colleague Pippa Norris found that these "long-term cultural shifts are important in bringing greater equality between women and men."[22] And with colleague Christian Welzel, Inglehart showed how democracies depend on a critical mass of self-expression values within the population. In other words, aspirations for self-expression support an egalitarian view of gender as well as a strong democracy.

Inglehart and Welzel themselves cited Maslow: "Aspirations for choice and self-expression are universal human aspirations: attaining them brings feelings of self-fulfillment, as Abraham Maslow pointed out long ago."[23] To explain how self-expression values lead to more democratic societies, they invoked Maslow's terminology: "As individual safety and autonomy reduce egocentrism, they increase homocentrism"–the concern for all humankind that Maslow saw in self-actualizing and self-transcendent people.[24]

"What happens to [people's] desires when there is plenty of bread?" Maslow asked. "At once other (and higher) needs emerge."[25] Inglehart and Welzel agreed: "Experiential factors, such as whether people feel that survival is secure or insecure, are at least equally important in shaping people's worldviews."[26]

Of course, this is a broad-brush explanation for the relationship between intrinsic growth and national change.[27] Reality isn't so pat, and there are any number of factors that affect individual nations. Nevertheless, the data show that on average and at large scale, changes in mass values correlate with national development. Ultimately, Inglehart and his colleagues have shown that something like Maslovian growth, some mass shift in aspirations, is tightly linked with modernization.

India's Tech-Sector Secret

In 2006 I visited Microsoft's sales office in Karachi. As I traveled around the city, I couldn't help but notice the contrast between Pakistan and India. In urban India, litter was everywhere, and migrant families crammed every inch of available space. In Karachi, the alleys between high-rise apartments were empty, and there were large tracts of open land without squatters. It seemed clean and orderly.

But looks can be deceiving. The people I met there saw things differently. They were envious of India's high-tech economy. They knew I had helped build a small part of it in Bangalore, so they all wanted to know: What was the secret to India's IT sector?

I was used to this question. In Brazil they wondered how India–historically much poorer than their country–was doing so well. In Kenya they asked what it would take to replicate India's success. In Sri Lanka they talked about leapfrogging, like India, from an agricultural economy directly to a service economy, with no industrial phase in between. These questions didn't always take into account the whole context of India, but they were insistent enough that I looked for answers.

Indian professionals explain their technology success in several ways. Many people point to 1991, when Prime Minister P. V. Narasimha Rao and his finance minister, future prime minister Manmohan Singh, opened up the country to foreign investment. The reforms they put in place unleashed the country's economic boom. Others note how Indian technology companies got a boost from the Y2K scare. Corporations worldwide worried that their computer systems would reach the year '00 and mistake it for 1900, not 2000. The fix–to upgrade to four-digit years–was technically straightforward but tedious. Indian companies filled the demand with low-cost engineering. Still others mention that educated Indians–unlike their Chinese peers–speak English, and so they can readily communicate with American firms.

But none of these explanations satisfied the Pakistani professionals I met. Pakistan, too, had been open to foreign investment since at least 1991. The Y2K crisis was worldwide, and Pakistani companies could have risen to the opportunity. The Pakistani elite also speak English.

The real secret, evident all around me in my Bangalore office, was India's decades-long cultivation of its brightest engineers through institutions such as the IITs. As anyone in Silicon Valley can tell you, IIT stands for the Indian Institute of Technology, a set of universities that graduate the country's top technical talent. They were inaugurated in 1953 by Prime Minister Jawaharlal Nehru, and today the acceptance rate at these government-subsidized schools is lower than that of Harvard's or MIT's. At a fiftieth-anniversary celebration in 2003, the CEOs of Amazon, Cisco, and Microsoft not only praised the institution for being "unique," "incredible," and "world class," but also thanked it for its global contribution.[28] Bill Gates said, "The computer industry has benefited greatly from the tradition of the IITs." He was right to be grateful–at Microsoft, as many as 20 percent of the employees are from India, many of them IIT alumni.

With so many graduates going abroad, you might wonder what the IITs are really doing for India. But their departure doesn't necessarily mean a permanent brain drain. AnnaLee Saxenian at Berkeley's School of Information found that, when the conditions are right, brain drain can turn into beneficial "brain circulation."[29] In her book *The New Argonauts*, she ascribed much of the success of the technology sector in India–as well as in Israel, Taiwan, and other countries–to exactly this phenomenon.

A perfect example of brain circulation is my former manager at Microsoft Research. P. Anandan graduated in the late 1970s from IIT Madras and left soon afterward for the United States to pursue a master's degree. A PhD followed, and then research jobs at Sarnoff Corporation and Microsoft. By the time I began working for him, he had lived happily in America for about twenty-five years. He was a classic case of Indian brain drain.

Until he wasn't. For as long as I had known him, Anandan had lobbied to support Indian computer science research. He had started an internship program for Indian students. He had made frequent trips to Indian universities. He had consolidated support from Indian executives at Microsoft. So one day, the management, going all the way up to Bill Gates, gave the green light for a research lab in India with Anandan as its founding director. A few months later, with me in tow, he moved back to a country he had not lived in for a quarter of a century.

This is Saxenian's theory in action. Anandan brought to Microsoft Research India all of the skills, attitudes, and social networks he had cultivated abroad. As we set up the lab, we tried to bring in the best of American research culture–academic quality, intellectual openness, minimal hierarchy, and unapologetic pride. As a result, we were able to attract some of the strongest computer scientists in the world, including quite a few who, like Anandan, were born and educated in India and had spent years abroad as researchers. (Of course, we merged all of this with a distinctly Indian culture, as well: Our employees were decidedly more open on a personal basis with each other than I have experienced elsewhere. There was tremendous acceptance of diverse thinking and processes. And we were masters of brinkmanship–event preparations miraculously came together at the last possible moment.)

The role of the IITs is mirrored by other reputable educational institutions in India, including the Indian Institute of Science, the University of Delhi, the Indian Institutes of Management, and the All India Institute of Medical Sciences. And technology-sector success is influencing India's other industries.

Where can we locate the true origins of the large-scale change that started with the Indian technology sector? Though it was undoubtedly enabled by foreign direct investment, high-caliber engineering education and internationally experienced managers were absolute preconditions. A single change in law wasn't enough. We can see this because we know that foreign direct investment in and of itself isn't sufficient. In India, economic benefits have gone disproportionately to the well educated–though the whole country lives under the same laws, recall that over 800 million people there still live on under $2 a day. And around the world, the lowering of investment barriers doesn't automatically result in a thriving IT sector. Pakistan, Brazil, Kenya, and Sri Lanka are all trying to various degrees, but unlike the passing of investment policy or the importing of hardware and software, talent takes at least a generation to mature. India's growth was predicated on decades spent building human capacity–some of it deliberately fostered by wise decisions in Nehru's generation, some of it built up as individuals pursued their aspirations abroad. The packaged intervention of finance reforms amplified national intrinsic growth.

It's also important that what is superficially a story of technology and economic growth is at least as much a story of changes in heart, mind, and will. If you visit newer tech companies in India, you can't help but notice how much they differ from the country's more established businesses. Sure, the buildings glitter with glass façades, and there are foosball tables in the hallways. But the bigger differences are in workplace culture and in the social norms that employees take for granted. For example, India's traditional industries tend to be run in a command-and-control manner. Your boss always knows best, and you do what he tells you to do. India's tech industry tends to be flatter. I saw this difference all the time: People we hired from outside of the tech sector took time to adjust, either because people didn't respond well to their barked orders, or because they were suddenly asked not to call their managers "Sir." Another difference is a reevaluation of what counts as career success. An earlier generation almost uniformly saw a stable job as the height of accomplishment. College graduates wanted to become doctors, lawyers, and government bureaucrats. Today, mere financial security has lost its sheen. The most ambitious IIT graduates are leaving large corporations to start their own companies. They're aspiring for greater esteem, achievement, and self-actualization.

Self-Actualizing Creative Class

Inglehart's analysis tapered off with the service sector, and that's where the sociologist Richard Florida picked things up. He extended the analysis with investigations of what he popularized as the *creative class*.[30] These are "scientists, engineers, artists, musicians, designers and knowledge-based professionals" who are "paid principally to do creative work for a living."[31]

The rise of the creative class is an international phenomenon led by developed-world cities. Florida estimated that in the United States the creative class grew from 3 million to 38 million people between 1900 and 2000, or from 10 percent to 30 percent of the population. A "Super-Creative Core" went from 2.5 percent in 1900 to 12 percent in 2000. But these figures underestimate the influence of the creative class, who, combined, command half the wages of the US workforce, control many of the most powerful institutions in the country, and design the form and content of the goods that consumers buy. In economically advanced European countries, the creative class makes up a similar share of the workforce.[32]

Among the creative class, self-actualization is a prominent theme. Florida wrote, "Creative workers do not merely move up the scale in Abraham Maslow's classic hierarchy of needs. Most are not very worried about meeting the basic needs of subsistence; they're already on the upper rungs of the ladder, where intrinsic rewards such as esteem and self-actualization are sought."[33]

Florida's creative class is less motivated by the lower levels of Maslow's hierarchy: Money represents not survival and security but "just a way to keep score," and "the best people in any field are motivated by passion."[34] Even esteem needs take a particular form. Florida found that those in the creative class who are born into money "no longer find true status in their wealth and thus try to downplay it." Instead they seek "the chance to win the esteem and recognition of others in the know."[35]

Thus, members of Florida's creative class work primarily because they love what they do, just as Maslow's self-actualizers do. The creative class values autonomy and diversity, just as self-actualizers do. Members of the creative class are confident of their ability to fulfill basic needs; when their jobs are threatened, their primary worry is not that they'll lose income, but that they'll have to settle for "just a job."

One graph in Florida's book captures the link between individual aspiration and national demographics. In 1900 agriculture was at its peak, employing nearly 40 percent of the workforce, but it was soon taken over by a reign of the working class, which usurped a similar chunk of the workforce between 1910 and 1960. Since the 1970s, the service class has been dominant, with about 45 percent of the workforce, but this dominance is now poised to fade as the creative, self-actualizing class continues its climb. These labor cycles are stark reminders that mass shifts in individual Maslovian development correlate with national socioeconomic growth.

The Chicken *and* the Egg

Inglehart and Florida's core claims stress that national changes correlate with changes in individual aspirations. You can't have one without the other.

If this seems obvious, it's worth recalling that social causes are suffused with attempts to impose supposedly wise outcomes from the outside without any attention to people's values and aspirations. Projects focus on bed nets, laptops, and improved seeds but with minimal focus on nurturing the people who will use them. Increased GDP is always cheered, even if it's caused by skyrocketing health-care costs, or a few rich people getting richer while everyone else languishes. Elections are expected to lead to democracy even in places where internal institutions and social norms are not ready for it. So many projects seek to improve a superficial condition rather than bringing about deeper change.

Institutions hastily built without a foundation of intrinsic growth have a nasty habit of evaporating. The democracies of Iraq and Afghanistan are good examples. Inglehart and Welzel launched a stinging attack on any notion that holding elections creates happier, more tolerant, or more freedom-loving people. Their data show that in the post-Soviet countries, the transition to democracy in the 1990s was followed by decreases in trust, tolerance, and happiness. This is significant, because "whether democracy takes root seems to depend on the strength of self-expression values far more than on simple habituation through living under democratic institutions." They concluded, "Even the best-designed institutions need a compatible mass culture."[36] Democratic packaged interventions only work when the underlying citizen values are conducive to them.

So which comes first–individual intrinsic growth, or societal intrinsic development?[37] The likely answer is that they cause each other. Individual growth tends to lead to national growth and vice versa. Socioeconomic growth and Maslovian development, like many strongly correlated social phenomena, are mutually reinforcing, at least up to a certain point of wealth. As people become richer, they aspire toward something more than mere survival. And as more people aspire to something beyond survival, society makes itself richer.

As long as individual and societal growth fortify each other, we should attend to the controllable causes of both. At an individual level, we should strive for intrinsic growth in ourselves and others. At a societal level, we should seek to establish institutions and policies that reflect and foster societal heart, mind, and will. And as for the political will to follow through, what is political will, other than mass intrinsic growth?

A Compassionate Class?

Earlier in this chapter, we encountered Narasimha, an example of an individual undergoing intrinsic growth who is a member of India's working class. Now I turn to children of the more comfortable classes in India, among whom another form of growth is taking place. They are more concerned with other aspirations.

Among the researchers I worked with in India were Nimmi, a housewife and mother who dusted off a degree in social anthropology to turn to the study of technology in urban slums; Saurabh, a talented computer scientist who thought about what he could do for India's underachieving school system; Aishwarya, a passionate development economist who balanced her academic calling and public service; and Indrani, a design researcher who wanted to coax more social impact out of her research.

They all came from relatively well-off families, and while their concerns were unique, they were largely indifferent to financial and social security. Instead they sought esteem, achievement, and self-actualization to various degrees and also felt the pull of self-transcendence. Viewed against the backdrop of their parents' generation, they exhibited a change in aspiration–they wanted to share what they had with less privileged communities.

If the creative class represents mass self-actualization, then it's tantalizing to imagine mass self-transcendence leading to a *compassionate class*. Imagine a country whose own needs are thoroughly met and whose dominant aspirations are world peace and prosperity. Such a country would let go of economic growth as its primary national goal, though it would still be able to meet the needs of its own citizens. It would be a nation of self-sufficient, altruistic people seeking to encourage intrinsic growth in themselves and others, and *not* in the self-serving or proselytizing way of a neocolonial White Man's Burden.

I believe there are signs that some strata of the world are already headed this way. In the United States, for example, the rate of charitable giving and volunteering is near all-time highs despite a small dip since the recession. There are over 1.5 million nonprofit organizations today, and each year, about 50,000 new ones are formed.[38] In 2011, 27 percent of American adults contributed 15.2 billion hours of volunteer time for various causes, a historical high.[39] Total charitable giving in 2013 was $335 billion, near the peak reached in 2007 just before the recession hit.[40]

These trends seem especially strong among younger Americans. In 2013, the Higher Education Research Institute at the University of California, Los Angeles, surveyed students and found that 71.8 percent of freshmen felt that "helping others" was an important part of life, the highest rate recorded since the late 1980s.[41] Frank Luntz, "the Nostradamus of pollsters," noted that "the 2020 Generation wants to do good as they do well."[42] Many of these students are children of well-to-do Gen-X'ers who grew up in a world that was not only financially secure but able to offer a certain amount of social status and self-actualization as a given.[43] Their parents were already engaged in jobs they loved, so, just as a previous generation assumed that financial security was assured, this generation assumes that a satisfying occupation is assured. What they're sensing now might be the early aspirations for self-transcendence.

David Bornstein, author of *How to Change the World*, cited similar trends elsewhere: Most of the 20,000 nongovernmental organizations operating in Bangladesh were begun in the past quarter century. France saw an average of 70,000 citizen groups established each year throughout the 1990s. During the same period, Brazil saw a 60 percent increase in the number of registered citizen groups. In that decade, the number of international citizen organizations grew from 6,000 to 26,000.[44] Lester Salamon and his colleagues at Johns Hopkins University found that "a major upsurge of organized, private, voluntary and nonprofit activity, has been under way around the world for the past thirty years or more."[45]

Some countries go further, making charitable goals central features of state policy. Sweden and Norway lead the world in official development assistance, with both countries providing over 1 percent of their gross national income. They are followed closely by Luxembourg, Denmark, and the Netherlands.[46] The same set of countries is at the top in terms of humanitarian aid contributions per citizen, with Luxembourg coming in at $114 per citizen per year in 2008, and Norway, Sweden, Ireland, and Denmark following.[47] It's no accident that these countries all register very high on both secular-rational and self-expression values in the World Values Survey. Perhaps they've saturated those scales and are moving into a more compassionate dimension.

I don't mean to suggest that any significant part of the world is transcending into some sage-like state of pure altruism.[48] Foreign aid programs have plenty of problems. And the desire for glory through activism can be just as strong as that for glory through wealth. Many activists put on a mantle of public service but still seek recognition or hero status.[49] There are, however, hopeful signs. The recent rise of the creative class in the developed world is something never before seen. Gender disparities around the world are shrinking. And as psychologist Steven Pinker cataloged in his tour de force *The Better Angels of Our Nature*, rates of human violence have fallen across the long run of civilization.[50] That more and more people are shifting their concerns to something other than corporate climbing, Wall Street riches, and selfish esteem suggests the dawn of a new aspiration.

Further growth is possible, but it's not assured. It's not clear that a broadly compassionate world could ever be a reality. Modern global civilization seems stuck in a form of self-actualization marked by consumption and personal achievement. But while a self-transcendent nation is not a sure thing, it is something to work toward. It's a dream worth believing in, a possible self-fulfilling prophecy, and a brighter aspiration for our future.

CHAPTER 10

Nurturing Change

Mentorship as a Social-Cause Paradigm

My parents were relaxed about academic achievement, but the "tiger mother" in them roared when it came to my learning the piano. They started me on lessons when I was five and didn't let me quit until I left home for college. I hated to practice, but my mother forced me to spend an hour a day at the piano–more in the days leading up to recitals or competitions.[1] She would accompany me to piano lessons. At home, she'd sit next to me, repeating my teacher's instructions.

"You may never become a professional pianist," my father said, "but you will learn useful things." At the time, it sounded to me like adult mumbo-jumbo, but looking back, I realize he was right. Today I'm at best a clumsy pianist, but I gained much sitting at the keyboard. I learned how to practice complex skills, memorize pages of information, find pleasure even in tedious tasks (like scales), find a place for creativity in repetition, work alone and work with others, handle the anxiety of public performance, interweave intellect and emotion for aesthetic goals, and so on. And I learned perhaps the most valuable lesson of any good education: that so much of life is rooted in both skill and habit, and that almost everything can be learned with sufficient practice. Music, of course, is hardly the only way to learn these things. Others learn similar lessons through sports, dance, visual arts, academics, computer programming–or any other pursuit that involves long-term training and engagement.

There are no shortcuts to mastery. There is no technology, no institution, no policy, no method that can turn a novice into a concert pianist overnight. Metronomes and tuners are important aids for learning. And technologies like radio and iPods allow you to "scale" music very easily, if merely playing it back is the goal. But you'd never call someone a musician just because they could flip the switch on a player piano. They have to be able to create music themselves. True virtuosity requires years of motivated effort.

If that's true for an individual, what would it take to create a whole nation of concert musicians? Venezuela offers a clue. In the 1950s and 1960s, its two biggest cities, Caracas and Maracaibo, were home to a few professional orchestras. These elite groups were funded by Venezuelan oil, but almost all their members were European or North American. José Antonio Abreu, a petroleum economist who played piano and violin, was dismayed by the lack of opportunity for Venezuelan musicians, so in 1975 he called together several friends to start a youth orchestra. On the first day, just eleven people showed up in a garage. But they were united by passion, practiced all day, and then went home and recruited others. On the second day, there were twenty-five. On the third, forty-six.

Within a month, there was an orchestra of seventy-five members, some of whom could barely play their instruments. Abreu held rehearsals for up to twelve hours a day. He asked older musicians to mentor younger ones. And soon they were performing for government ministers and dignitaries, such as Mexican president Luis Echeverría Álvarez, who was so impressed that he invited them to perform in his country. Building on this early success, Abreu took the orchestra to festivals abroad. Quickly, it won international acclaim.[2]

Over time Abreu was joined by others who shared his vision. They established facilities across the country, typically in poorer neighborhoods. They secured funding from public sources: the Venezuelan government, the World Bank, the Inter-American Development Bank, UNICEF, and others. The effort grew amid major political transitions. Today there are 285 *núcleos* across Venezuela, where over 400,000 students spend much of their free time. They practice and rehearse, hoping to gain entry to one of the network's thirty-one professional symphony orchestras.[3] You can walk through barrios and hear the music of Bach and Mahler wafting through the air. The country has so embraced the phenomenon that in rural villages, families give their cows names like Mozart and Beethoven. The nation's best musicians form the Simón Bolívar Youth Orchestra, which has performed at Carnegie Hall, the Kennedy Center, and major venues in Berlin, London, Vienna, Tokyo, Beijing, and elsewhere. Rave reviews come in even from the most curmudgeonly critics. In 2009 the orchestra's twenty-eight-year-old conductor, Gustavo Dudamel, was tapped to become music director of the Los Angeles Philharmonic–a Venezuelan wunderkind heading one of the world's top orchestras. He has been called "the next Leonard Bernstein."

Abreu named his organization Fundación del Estado para el Sistema Nacional de las Orquestas Juveniles e Infantiles de Venezuela, or El Sistema for short. It's not a system, though; it's a social movement. El Sistema now includes a degree-granting conservatory, a network of academies devoted to specific instruments, courses in jazz and Venezuelan folk music, apprenticeships for luthiers, curricula for special-needs children, preschool classes, and even programs for new mothers and their babies. El Sistema groups have also branched out into about fifty other countries.

If you came across a country without music, you wouldn't think that it could be turned into a musical paradise through any packaged intervention. Yes, you'd need technologies such as violins, and maybe MP3 players, too. You'd need spaces where groups could rehearse. You'd need funding and perhaps a law or mandate in support. But none of that would ensure practice or legislate virtuosity. No packaged intervention can make a society genuinely musical without something more–something dramatically more, like the single-minded devotion of capable leaders, public support and generous sponsors, a network of committed musicians and administrators, a horde of music-loving children nurtured one by one, a society rallying with encouragement–and all of it spanning generations.

So many of our ideas about packaged interventions would seem laughably one-dimensional if we saw social change less as fixing a broken machine, and more like the cultivation of an orchestra. Social progress requires a change in people, and that can take years before it manifests as measurable behavior.[4] It's an obvious story, but it's also a story that's increasingly overlooked, drowned out by the flashy noise of cheap, packaged, blingy quick fixes. Visible events, such as the Egyptian Revolution and the legalization of same-sex marriage in America, happen so quickly that it's easy to delude ourselves into thinking they were made possible by Facebook or by a single law or ruling. But to think that these are momentary events without a long history is to think that a symphony is just a flash mob of random people who decided on a whim to pick up instruments and play a sequence of sounds just so. The reality, of course, is that every note played on stage resonates with years, if not decades, of individual, collective, and intergenerational effort. There are no technocratic shortcuts.

But if there are no shortcuts, what can those of us who want to contribute to social causes do? This final chapter offers ideas for fostering intrinsic growth in others. If technology and packaged interventions are amplifiers of human forces, then the most important question in a world already rich with technocratic devices is how to nurture intrinsic growth.

Of Geeks and Gurus

I sat transfixed by Jayshree Diggi, a radiant thirty-year-old woman. She represented about forty women from two self-help groups who sat on the cement floor of a nondescript square building in eastern Jharkhand. In rapid Hindi, she explained how Lonjo, her village, had begun to incorporate a superior method of rice growing, as taught in Digital Green's videos. Diggi presented maps of the village and rattled off statistics about each family's gains in rice yield. Though some of the women in the audience were twice Diggi's age, they seemed to look up to her and rally around her.

That was July 2011. Several of us from the board of Digital Green had come to Jharkhand to observe the staff's work in rural India. We wanted to understand our partnership with Pradan, a pioneering non-governmental organization that works with about 350,000 families in seven of India's poorest states. Their programs span agriculture and health care, infrastructure, and local governance, and partner communities see diverse outcomes. Some have doubled or tripled their incomes. Others have cleaner water and fewer illnesses. Some take pride in young leaders such as Diggi. Many have built working ties with local governments.[5] Pradan strives to awaken poor rural communities in India to the possibility of their socioeconomic development.

Abdul Mannan Choudhury is the senior Pradan staff member for Lonjo. He says that when Pradan first began working there, Diggi was a quiet new mother who knew little about modern farming, finance, or politics. She was like many other Indian villagers–born in remote rural areas, raised with little formal education, used to a life of hard agricultural labor, married off young, ignorant of life outside the village, deferential to authority, and discouraged from having thoughts of their own.

But the person Diggi had become was nothing like that. She and her fellow self-help group members not only knew the intricacies of the new rice-planting techniques, but also took out loans from banks and were active in the local *panchayat* village administration. One woman in the group confessed that she had at first felt intimidated by bank employees, but was now able to deal with them. Diggi told us how she and the other women gradually inserted themselves into village politics. Today, over half of the participants in village meetings are women. The women also meet routinely with government officials. Diggi volunteered that she was proud of her work, but that all of it was ultimately for her children and for her village. She wanted to assure the next generation a solid education.

Those of us on that trip were ostensibly there to see Digital Green's video production process. We were also curious about the agricultural techniques recommended by Pradan. In other words, we were there to see the geeky packaged interventions. But as we interacted with the women in the self-help groups, it became clear that *they* were the real story. Diggi mesmerized us with her intelligence and charisma.

Choudhury mentioned that over the years, he had witnessed her "enhance her skill of interaction and presentation in the community, and increased her confidence by manifold."

So, how did Diggi become the person that she is? It wasn't the rice-planting techniques she learned in Digital Green videos, at least not in themselves. It wasn't the loans she took out with her self-help group. It wasn't the votes she logged at her local village meetings. Something much more than behavior change enabled by technologies, coerced through laws, manipulated via incentives, or exchanged in a marketplace was at play. Pradan's unique engagement with Diggi and her self-help group had catalyzed an intrinsic transformation that was not unlike El Sistema's transformation of musical novices into orchestras. What changed everything for Diggi was Pradan's *mentorship*.[6]

The Model of Mentorship

In mentorship, one party helps another gain the capacity to achieve the latter's aspirations, and both attain greater intrinsic growth. Mentors and mentees can be individuals, groups, or whole nations. Over a span of thirty years, Pradan has developed a set of tactics for how it engages with villages. By going through them, we can understand the elements of good mentorship. I use Pradan as an example because I happen to know it well, and because it is remarkably good at what it does. But there are many other organizations, often unsung and underfunded, whose primary model is mentorship, whether they call it that or not. Shining a light on the components of good mentorship will make organizations engaged in it more visible and help other organizations become more effective.

Building Relationships with the Mentee in the Driver's Seat

After Pradan identifies a community to work with, a staff member spends a few days there each month for several months to establish rapport with residents. Choudhury's predecessor in Lonjo was the first one to work with the village. He began by seeking out residents to speak with, helping them with chores, and otherwise spending time with them. A relationship based on trust is essential to good mentoring, but good relationships may take a while to establish.

Mentorship also explicitly acknowledges any status difference between the parties. In the case of Pradan and the villages it engages, this happens naturally. The disparities are impossible to hide. Pradan sends its people where they know they can help. When staff members arrived in Lonjo, they met villagers who were poor, passive, and poorly educated. In contrast, Pradan employees are typically university graduates who speak both Hindi and English. They ride in on motorcycles that most villagers can't afford, and they have a young, urban bearing that many villagers will only have seen on television. Acknowledging the disparity in status is the basis for vigilance against abuse and exploitation.[7] Anirban Ghose, Pradan's program director and former executive director, says, "Mentee communities are vulnerable, and it is our job to engage with a spirit of 'trusteeship.'"

If the relationship is weak, or if the village is uninterested in what Pradan has to offer, the relationship ends. Potential mentees have the right to decline mentorship. Unlike with the imposition of law or projects run without community consultation, both sides must freely choose to be in a genuine mentoring relationship. Conversely, mentors may avoid activities that they don't believe in or refuse to work with potential mentees who don't seem sincere. Pradan sometimes leaves villages that don't appreciate its intentions or its expertise. (Walking away like this is very different from a related but unfortunate practice common in international development–the coerced deal: "We will provide economic advice and $1 billion in low-interest loans, if you lower import tariffs." Or, more generally, "We will provide X if you do Y." That's not mentorship; that's extortion.)

Mentees should own the agenda and set the terms of engagement. Mentees need to stand firm about their own aspirations and avoid second-guessing their mentors. And mentors must resist the temptation to impose their worldview onto mentees. Developing these attitudes can itself take effort, and it's part of the work.

With the mentee in charge of the agenda, though, mentors must be ready for surprises. I once heard a story from A. V. M. Sahni, a retired Indian Air Force lieutenant who worked with the nonprofit Development Alternatives. He was a proponent of check dams–small dams that slow the flow of rivers and allow adjacent fields more time to absorb water. Check dams turn cracked dirt into moist, loamy soil even in the dry season. After Sahni coached a few communities in and around the central Indian city of Jhansi in how to build check dams, residents found they could easily grow two crops per year, whereas earlier they were limited to one. Overnight, they doubled their annual incomes, and they were overjoyed. Sahni was more ambitious, though. He felt they could get three crops in. It would only take a slight adjustment in their planting schedules. As he suggested the change, however, the farmers rebelled: "Who is this old man, who has nothing better to do than to walk from village to village trying to make us do more work?" Sahni told me this story with a smile, and said, "You know, there's no point pushing people beyond what they want for themselves."

Waking Aspirations

Once a relationship has been established, Pradan recommends that the village form self-help groups that meet regularly. Diggi joined one of two that were formed in Lonjo in 2002. Pradan's work here is extensive, as Choudhury explained:

> In the very beginning, there was a concept-seeding meeting. That was followed by an exposure visit to some of the female and male members of an existing SHG [self-help group] promoted by Pradan in the neighboring village of Edalbera, who oriented them about the nitty-gritty of SHGs. There was interaction with existing SHG members who discussed an SHG's operational principles, its by-laws, and its benefits for a member. After the exposure visit, another meeting was conducted to hear if village residents were interested in membership. In the subsequent meeting, an SHG was formed.

Pradan's gentle yet persistent facilitation results in a vibrant esprit de corps within the self-help groups. Today there are thirteen self-help groups in Lonjo and more than 22,000 across India.[8]

Once formed, an SHG's first task is to articulate its aspirations. But in communities used to the losing side of justice, aspirations might need to be awakened. "The greatest challenge in our work is not corrupt bureaucrats, not insufficient government programs, and not internal strife within villages and families," Ghose says, "though all of these problems exist." The greatest challenge is overcoming the villagers' capitulation to the status quo. They ask, "Even if God Himself were to appear here things are not going to change, so what chance do I have?" Pradan therefore provides gentle, frequent encouragement, building on whatever strengths villagers already have. Eventually communities are able to identify the fragile hopes hidden under layers of protection against disappointment.[9] "We'd like to be less dependent on moneylenders." "How can we wean our husbands off of drinking?" "We'd like to improve our rice yields so that we have more money to spend."

This process could be called an *aspiration assessment*, in contrast to the needs assessment so many social activists are trained to perform. The intent behind the needs assessment is a good one–efforts should obviously go toward what communities actually need, not what we, as outsiders, think they need. In practice, though, it's nearly impossible not to project needs onto others. Richer people believe that their poorer counterparts need better health care. Citizens of democracies are certain that others need political freedom. University graduates assume that everyone needs higher education. For most readers, these are without doubt essential, but for other people in other circumstances, these "needs" may come second or third to more insistent desires.

In contrast, people are deeply vested in their aspirations. One example comes from the mobile industry. Research finds that even very poor people will work hard to pay for high-cost, high-tech smartphones.[10] The underlying aspiration is for status and cosmopolitan sophistication. Mentorship would help people meet the substance of the aspiration. It would help people increase their knowledge, self-confidence, earning power, or social influence. (In contrast, a technology-centric approach would give away free cellphones.) Aspirations, more than needs, stretch heart, mind, and will.

For Pradan, the foremost question is, "What does the community *want* to do?" Though it offers suggestions, Pradan is careful not to impose too many of its own ideas.

Driving Intrinsic Growth

The point of aspirations is to drive intrinsic growth, so once aspirations are articulated, Pradan brings in technical experts. These experts teach skills, help obtain resources, and connect the SHGs to relevant institutions. According to Choudhury:

> [Pradan first provided] training about how SHGs work. We walked them through savings programs and microcredit. We explained their legal rights and entitlements. Then, we discussed social evils like witchcraft, alcoholism, and domestic violence. A few select SHG members, including Diggi, were trained on legal actions open to them, gender issues, etc. They were informed about different natural resources available in their community. Land-and water-based works were conducted in partnership with the Jharkhand Tribal Development Society, who provided funds for irrigation infrastructure. Most recently, Lonjo was introduced to new rice varieties, and we worked on improved practices around rice cultivation.

Throughout, the focus is on increasing the community's capacity. Pradan provides assistance and coaching so that the self-help groups are able to meet their aspirations on their own. People such as Diggi are direct recipients of Pradan's tutelage.

Pradan avoids doing a village's work itself. Instead it facilitates the process by which villagers learn to achieve the goal for themselves. Mentoring may involve demonstrations, but Pradan doesn't implement anything on behalf of villages on an ongoing basis. If Pradan attends to Lonjo's agricultural output, it's primarily because it signals Lonjo's own increased learning. The increased output means that Pradan's mentorship is working. Mentors provide encouragement and inspiration, even occasional pressure, but they don't harass or dictate milestones.

Mentorship thus focuses on fostering knowledge, skills, social networks, and other forms of individual and communal wisdom. Pradan's main gifts to a community are its staff's time, energy, and expertise, as opposed to money, food, equipment, infrastructure, or technology. Pradan does obtain grants and donations for packaged interventions, but all such gifts are in service of the village learning something new. (Pradan's long-term ideal is for villages to learn to secure their own funding.) Pradan's relative disengagement from material provision has many implications, but one of them is that mentorship is relatively resistant to corruption. A million dollars of aid dropped in a government bank account easily slips through the greasy cracks of corruption, but $1 million of time from mentors is much harder to tuck into a pocket. That said, mentorship must be effective and appropriately priced. There's no value in lining the pockets of overpaid consultants claiming to be mentors but doing little other than filing reports and securing their next project.

Focusing on mentee aspirations avoids the problems of labeling people or blaming the victim. The hierarchy of aspirations is meant primarily to map the landscape of human progress. It demonstrates that summits exist and that it's possible, in theory, to reach them. But the map doesn't provide specific routes, which depend on where people start their journey as well as their particular preferences and capacities. Where people are on the mountain matters less than that they climb. Knowing which way is up is all that matters, and *up* is indicated by aspiration.

In our role as mentors, we should foster the capacity for people to choose and take their own aspirational steps. We don't have to decide the step for them. If there is an end goal to mentorship, it's to reach a point where mentees can achieve their goals on their own. This is exactly why Pradan puts so much focus on fostering autonomy in self-help groups.[11]

Teaching and Parenting

Despite what I've said above, there are cases where the mentee's aspirations are either unclear or unformed. This is usually the case with children, who require a special kind of mentorship–parenting and direct instruction.

But I have also met adults who have had so little say in their own lives that they seem terrified of making the smallest decisions for themselves. In Pradan's communities there are, of course, women who never become like Diggi. They're more comfortable following their husbands or their peers. Still, they have some urge to grow. Imparting relevant skills nurtures not just knowledge but confidence in the ability to learn. For such people, teaching can be an effective form of mentorship, as long as the focus remains on intrinsic growth. More generally, there are a whole line of educational categories that deserve greater attention in social causes: primary, secondary, and tertiary education, of course, but also vocational education, job awareness, soft-skills education, confidence-building exercises, artistic or athletic skill development, leadership and management training, community building, organizational development, so-called capacity building, and so on. Depending on the context, some of these may fit best with individual or group aspirations.

And That's Not All

Intrinsic growth, and therefore mentorship, takes time. Pradan commits to villages for years, sometimes decades. Often it stays until self-help groups are fully autonomous and there is nothing left to teach or support. Mentors provide encouragement and inspiration, even occasional pressure, but they don't harass or dictate milestones.

Pradan's sole objective is benefit for the village. Mentorship is fundamentally different from the transactional processes of business, trade, quid-pro-quo politics, and other forms of exchange. In some transactional relationships, a lower-status person performs some work for a higher-status person, or pays them directly in exchange for their advice and help.[12] This may look like mentorship, but the vested interest in the relationship can lead to exploitation. Direct, tangible benefits for the higher-status party conflict with the true goals of mentorship. Any material exchanges should be undertaken with care, transparency, and even suspicion.

Of course, if there were absolutely nothing in it for mentors, mentorship would be a hard sell. But mentoring is both gratifying and empowering for the mentor. I have volunteered for several programs where there was an oversupply of mentors, and we competed with each other for the attention of prospective mentees. When done well, mentoring is a deeply caring, empathetic activity. If Aristotle was right that "we become just by doing what is just, temperate by doing what is temperate, and brave by doing brave deeds,"[13] then it seems likely that through mentoring, we become more giving, self-transcendent individuals.

Mentorship differs significantly from attempts to "roll out" or "scale up" packaged interventions. If helping people reach their personal aspirations is the focus, then agendas have to adapt accordingly. Pradan, for example, is prepared with a range of packaged interventions, but it calls on them only when village communities request them. The focus is not on pushing packaged interventions onto every community, but on matching packaged interventions with a community's aspirations. "We do not see ourselves as a service delivery organization, but rather [as being] in the business of rekindling hope and fostering processes," Ghose says. "We help develop knowledge, skills, and [social] linkages that help poor communities participate on an equal footing in society."

Lastly, it's important that the emphasis be placed on *good* mentorship, not just the trappings of mentorship, or mentorship as a trendy buzzword. There are any number of ways in which mentorship could go sour. Just because some mentorship efforts, possibly under other names, such as "capacity building," have failed or fallen out of favor doesn't mean that mentorship couldn't be done better.

Mentorship is not earthshaking conceptually. Yet, in part because it is rarely held up as a model in social causes, it has often been neglected by theorists, policymakers, and donor organizations.[14] Mentorship, though, works well as an overarching framework that avoids the problems of top-down authority, benevolent paternalism, or pretended equality.

What's Not Mentorship

Mentorship differs from other approaches to social causes, such as coercion, manipulation, charity, and trade. Coercion involves physical or legal force to constrain or command action. Examples include military force for regime change and China's one-child law to limit overpopulation. Manipulation is in fashion among technocrats as a way to induce "voluntary" choices through extrinsic incentives. Paying families to send their children to school through conditional cash transfers and imposing sin taxes on unhealthy goods are types of manipulation.[15] Charity, in the form of handouts without attempts to nurture people's capacity, can be helpful in emergency situations, but, when provided without reflection, it can become a crutch that stunts growth. And trade is often considered an unqualified good, but it easily devolves to exploitation when exchanges happen between parties of unequal power or wealth. Think blood diamonds and Nigerian oil.

Mentorship is most studied in the business world, where it is sometimes contrasted with coaching and managing, both of which share mentorship's acceptance of status disparity but differ from mentorship in important ways.[16] In certain definitions of coaching, coaches are content-free sounding walls and program managers: They introduce no technical knowledge to the relationship.[17] Unlike a coach, a mentor often brings relevant expertise and resources. Managing, meanwhile, is the art of causing others to perform constructive work, usually toward the manager's goals. The emphasis is on achieving the goals of the higher-status entity, reversing the objective of a mentoring relationship.

Any of the options above might be strategically used in a broader context of mentorship. I'm not recommending mentoring in opposition to other models, but mentorship as the overarching framework, as well as a conscious shift toward more mentorship. Mentorship allows for a judicious switching between modes. The broader goal of intrinsic growth might require a discerning application of other models, but always in service of the mentee's aspirations and intrinsic growth. Mentorship's long-term hope is the maturation of an independent peer or partner.

From External Packaging to Intrinsic Growth

Despite everything I've said about mentorship, it would be a shame to overlook technological amplification when it's appropriate. Lonjo took advantage of new seed varieties, savings schemes, and other interventions as it worked with Pradan. Packaged interventions are important–a wise person or a wise society will apply the right ones in discerning ways. But even in their application, there's always an opportunity to nurture intrinsic growth.

As we saw, Digital Green uses how-to videos of farming techniques as a teaching aid to persuade farmers to take up effective agricultural practices. To this end, it has an extensive online video library, maintains a data analytics system to track performance, and uses handheld video cameras and pico-projectors–portable projectors about the size of an ice-cream sandwich that can project digital video onto a wall in any dark room. So technology is a key part of its work.

But while Digital Green's ultimate goal is to help raise farmer yield, income, and welfare, its focus is resolutely on building human capacity. It does this in several ways. To begin with, Digital Green's videos and pedagogy are about nurturing knowledge and capacity among farmers. The videos are carefully curated for local relevance (to crop, season, climate, and geography), and they are ordered in such a way that content with the quickest return and most visible benefits are shown first. That helps build farmers' self-confidence and appetite for more knowledge. Next, Digital Green works mainly by teaching its partners its methodology. With Pradan, Digital Green began by directly overseeing the video processes in villages, but over time, the practice was taught to Pradan's staff, who in turn provide training to interested self-help groups.

So there is a packaged intervention here, whether you think of it as the how-to videos or the Digital Green methodology as a whole. But facilitating intrinsic growth–either of farmers or of partner staff members–is the ultimate goal. Again, it's not about the packaged intervention.

Anytime there's an intention to provide a packaged intervention, there's an opportunity to mentor. Most people agree that if you give a man a fish, he eats for a day, while if you teach him how to fish, he eats for a lifetime. But as we set aside turbo-charged, heat-seeking, robotic fishing poles to solve other people's problems, we could do more than teach people to fish. We could also support instructors to teach fishing, encourage entrepreneurs to manufacture fishing equipment, promote policymakers who can run well-regulated fish markets, nurture universities to do ichthyological research, cheer nations toward sustainable fishing, and on and on. That way other societies can foster their strengths, so that ours are less and less needed.

Evidence in agricultural development suggests that programs do well when they contribute to the growth of farmers and agricultural extension organizations. Studies attest to Farmer Field Schools, in which extension agents show farmers practical techniques in the field.[18] The higher crop yield associated with the Green Revolution in South Asia was due at least as much to ramped-up extension efforts as to the introduction of new seed varieties (putting aside questions of its negative effects on soil and agricultural systems).[19] Today China is among the few countries that continue to place faith in extension, deploying a force of nearly 800,000 agents unmatched elsewhere in quality or quantity. As a result, Chinese farmers are agile and productive.[20]

A packaged intervention can be thought of as a concrete focus for the sake of fostering intrinsic growth. It's great that Diggi's family earns more income today as a result of better agriculture. But what's truly remarkable–and self-sustaining–is her transformation into a confident community leader. For her, agriculture was a convenient focus that enabled more abstract learning.

Any packaged intervention and any activity can become a platform for fostering intrinsic growth. My friend Deogratias Niyizonkiza, founder of Village Health Works, uses health-care interventions as a starting point for mentoring a Burundian village to improve its own schools, food security, and livelihood.[21] I-TECH head Ann Downer once told me about K. Sethu, a taxi driver with a grade-school education whom she mentored into becoming the facilities manager of I-TECH's growing India office.[22] And, as El Sistema founder Abreu said, "An orchestra is first and foremost a way to encourage better human development." To him, music "is a primary instrument for the development of individuals and societies."[23]

Nurturing wisdom takes time, and intrinsic growth happens gradually. Unconfident single-crop subsistence planters are unlikely to transform overnight into cash-crop superfarmers who sell wicker furniture in the offseason. Struggling farmers might need relief before extension, and extension before diversification. This doesn't mean that people must be constrained to learning one thing at a time; it just means that great care must be taken to tailor programs for the context, for people's aspirations, and for their level of intrinsic growth.

It's Mentorship All the Way Down

Mentorship can be applied at various levels.

Pradan's leaders initially fell into a routine where frontline staff followed orders handed down from the leadership. But cofounder Deep Joshi quickly realized that it made no sense to expect self-confident decision-making from villagers when his own staff had to blindly obey orders from above. It was a philosophical contradiction with practical implications: Rural communities needed to see Pradan staff as role models.

So Joshi thought as hard about how he should run his organization as he did about what it should accomplish. He wanted the nonprofit to get used to "inducting, nurturing, and developing professional development workers who are . . . self-regulating, and continuously seek excellence in their tasks."[24] Over time, Pradan's senior leadership has developed a way of working that tries to do this. All staff members, for example, undergo a one-year apprenticeship that includes two weeks of immersion in the home of a poor rural family. During that time, they are effectively cut off from middle-class comforts. Most emerge with greater self-knowledge, a better sense for the hardships faced by poor families, and strengthened resolve to tackle them. In their work, they are given a lot of responsibility and latitude. Employees receive continuing training. Senior staff members join Pradan's governing council and act as mentors more than bosses. Some eventually move on to start their own nonprofit organizations. The staff becomes increasingly independent and autonomous, just as the communities they work with are expected to do.

Thus Pradan has multiple layers of impact: the staff cultivates intrinsic growth in the rural communities it works with; management cultivates intrinsic growth among staff; and leadership cultivates intrinsic growth within.

The Long, Hard Road

In *Memorabilia*, Xenophon, a student of Socrates, relates the story of a young Hercules, who meets two women at a crossroads. One offers him a life of ease and pleasure without notable achievement, the other a long, hard road that leads to "the most blessed happiness." Hercules, of course, chooses the latter.

Those of us who care about social change have a similar choice to make. The long, hard road focuses on mentorship, aspirations, and intrinsic growth, which are difficult to support in a technocratic world. They're not easy to measure. They resist quick scale-up. They're fraught with questions of values. They don't glisten with innovation. They violate all of the Tech Commandments.

Mentorship and intrinsic growth won't solve all of the world's problems. But if packaged interventions amplify human forces, then in a world already full of incredible technologies and brilliant technocratic ideas, what we need much more of is heart, mind, and will.[25]

Technocratic forces take care of themselves. We don't need to push for mobile services to reach every corner of the world–telecommunication companies have already done that. We don't need to worry about spreading profitable loans–eager banks will do so. We don't need to spark democratic protests–frustrated citizens will rebel on their own. Packaged interventions will thrive, with or without our help, because the easily replicable parts have built-in rewards.

What doesn't thrive on its own is quality education for all; the returns arrive after long delay and are hard to measure. Or economic on-ramps for hopeful strivers; there's no dollar value on aspiration. Or stronger communities and organizations; they require expensive person-to-person interaction. Or compassion among the rich and powerful; they can get along without it. In other words, we should focus on those goals for which technology and technocracy are ill suited: serving poor communities, educating the less educated, reforming dysfunctional institutions, organizing marginalized groups, preparing for long-term crises, encouraging self-transcendence, and eliciting responsiveness from those with power.

There are ready incentives for technocratic goals. What's left to be done is by nature difficult.[26] So, for anyone wanting balanced progress, for anyone with self-transcendent motivations, for anyone genuinely seeking social change, the most meaningful efforts are those not boosted by technocratic values. Packaged interventions are relatively easy. Nurturing individual and collective heart, mind, and will is hard. What we need is more people taking the long, hard road.

Conclusion

When I was fifteen years old, I won the egg-drop contest at my high school, the American School in Japan. The goal was to design the smallest, lightest contraption that would protect an egg in a fall from the school's water tower. My device nested the egg in a cardboard tube attached to a tissue-paper parachute. It would, I hoped, be my first taste of geek stardom.

My physics teacher, Mr. O'Leary, offered a hearty congratulations, and my classmates teased me out of envy. What I remember most, though, was that my victory went unmentioned in the next morning's public announcements. Our principal regularly played up sports team triumphs and drama club events, so why didn't a feat of engineering merit acknowledgment? It stung.

That night I thought about why I cared, and the sting gave way to curiosity. I had enjoyed designing the parachute and testing it off my eighth-floor balcony. My egg survived, and I could take pride in that. My self-image as a science whiz was preserved. So what did it matter if others knew? It seemed silly and vain to want more recognition.

I still think of that day as the dawn of my adulthood because I realized then that I was driven by powerful subconscious aspirations:

I sought certain kinds of achievement, and I wanted accolades. And while I knew at some level that it was better not to care about public esteem, the aspiration ran deep–I couldn't reason myself out of it.

The Life You Can Save

The philosopher Peter Singer opens his book *The Life You Can Save* with one of his favorite thought experiments.[1] Imagine you're on your way to work when you spot a young child drowning in a pond, but no one is around to save her except for you. Rescuing the child would require you to wade into the water, ruining your new shoes and making you late for work. What do you do? Of course, you would save the child. Weighed against her life, time and cost are nothing.

Singer then asks us to consider a real situation. Every day, thousands of children around the world die of various causes. Many of the deaths could be easily prevented for the price of new shoes. Measles, for example, kills about three hundred people per day, most under the age of five, yet the American Red Cross says that every dollar you donate is enough to vaccinate one child.[2] Most of us could easily afford a dollar a day by cutting back on coffee or choosing a cheaper mobile plan. Some of us could absorb the cost with no change in lifestyle. So why aren't we saving these dying children?

By juxtaposing the two situations, Singer argues that it's indefensible that we allow such tragedies. His point is compelling. Innovations for Poverty Action, a nonprofit that Singer endorses, recently received a donation accompanied by a note revealing the inner tension. It read, "Damn you, Peter Singer!" But for every such donor, there are hundreds, if not thousands, who follow the thought experiment and never write a check. When I read about Singer's drowning girl, my first thought was that I already made annual donations to several causes. Though I agreed with his reasoning, and though I could surely afford to give more, I didn't reach for my wallet. Why was that?

A slightly different hypothetical gets us closer to the truth: Imagine that you saved one child from drowning a couple of days ago. You promptly bought a new pair of shoes to replace your waterlogged loafers. Then, yesterday, you saw two children in the pond. You saved them both. More shoes. This morning, by some freakish coincidence, there were three drowning children. You saved all of them, too. But that's a lot of shoes to ruin in a week, and you've been late for work three days in a row. You're worried about tomorrow and the day after. What if, every day, more children needed saving? You doubt you can keep it up.

This is much more like the situation we actually face. Singer cites 27,000 children dying of preventable illness every day, or about 10 million a year. Most of us will happily save one child for a few bucks, but few of us will save all the children we possibly can on an ongoing basis. That would mean a commitment of time and money we're not ready to make. I'm quite happy to give up 0.1 percent of my annual income, or 1 percent, or 10 percent, or maybe even 20 percent. But 50 percent, 75 percent, or 90 percent?

In other words, the abstract good conflicts with my selfish desires. I give less than I could, consume more than I need, and spend time on activities such as writing this book–which, as much as I hope it serves a positive purpose, is also a bid for self-serving esteem. Even if I put aside guilt, shame, and every other self-admonition, the stark fact is that I'm no saint. I'm unable to be as kind as I know I should be. And that's the crux. Knowing isn't enough–I also have to become someone who can better execute what he knows.

Technocrats extol technology and knowledge and intelligence, but positive social change requires a lot more. Millions of people in the world today live satisfying lives envied by the rest. That means that we already have the knowledge we need for well-being. As foreign-aid critic William Easterly wrote, the technocratic illusion is to think that we suffer from a "shortage of expertise."[3] What we have instead is either a shortage of caring or a shortage of capable follow-through. Discernment is just the beginning. We also need superior intention and greater self-control. The question that Singer's drowning child poses is less about whether to save a child, or even which packaged interventions would save the most children, than about how we become the kind of people who can, and will, save more children.

This book's core thesis–that we should see social situations less as problems to be solved and more as people and institutions to be nurtured–has been raised by many others in different contexts. To various degrees, it aligns with Aristotle and Confucius and their intellectual descendants,[4] with advocates for health systems in public health, with the social worker's idea of social development, with Easterly's problem-solving systems in international development, with Evgeny Morozov's polemic against technological solutionism, with Diane Ravitch and David L. Kirp's critique of quick-fix approaches to American education, with 1980s communitarianism and its interleaving of public and private values, with the "institutional turn" in a range of social sciences, and with the thinking behind any number of feet-on-the-ground organizations that work primarily to foster people, organizations, communities, and nations.[5] I hope to have persuaded you that they're all united by a single theme. The lessons of one domain apply to the others. The difficulty of eradicating polio despite potent vaccines is congruent with the challenge of providing high-quality education in a world of digital devices, which is congruent with the obstinacy of prejudice in the face of civil rights laws, which is congruent with the obstacles to democracy despite elections, which is congruent with inaction on climate change in spite of green technologies. In the twenty-first century, we have plenty of packaged interventions. What we need more of are the right kinds of heart, mind, and will.

Grow Thyself

James Madison said, "To suppose that any form of government will secure liberty or happiness without any virtue in the people, is a chimerical idea."[6] Another American president, Abraham Lincoln, put it this way: "With public sentiment, nothing can fail; without it, nothing can succeed."[7] What both men were gesturing at is that our own intrinsic growth is essential. In fact, it's the most meaningful, sustainable, scalable, and cost-effective investment we can make.

We have much more control over ourselves than we do over others (however little it may feel at times), so it's a natural place to start.[8] And as we have seen, the societal effects of one person pursuing their aspirations can be significant. Patrick Awuah is just one man, but his growth ushered in new lives for hundreds of people, including Isaac Tuggun and Regina Agyare. And each of them is now spreading that impact to others. Some of us have influence on the lives of ten people, others on a thousand people, and still others on millions of people. Our intrinsic growth is vital because its impact is multiplied by the scale of our influence. The history of social activism is full of tales of bad intentions, poor discernment, and inadequate self-control among leaders causing nasty consequences downstream.[9] The closer we are to being the best versions of ourselves, the better the outcome for everyone we touch.

That holds among nations, too. When I lived in Bangalore, I was surprised by the degree of ecological awareness I saw among India's middle and upper classes. Companies advertised green products. Protesters marched against cutting trees to widen roads. Some of my friends started waste separation schemes in their neighborhoods. Compared with the timelines of most developed nations, India's environmental efforts have taken root early.[10] On a per-person basis, the country's economy is more than half a century behind that of the United States, but its green aspirations are similar. That's not to say that India is a haven–half of the world's twenty worst cities ranked by air pollution are in India–but some part of the country is planning for future environmental well-being.[11]

Many factors have contributed to this trend, ranging from India's own traditions of sustainability to the pressures of industrialization. But there's another factor to consider: As members of India's diaspora return, they bring sensibilities they learned elsewhere, including those supportive of ecological activism. India often looks to the West for cues, and it is not alone in this respect. Much of the developing world sees the materially developed one as a role model.

But if the developed world is to be a good model, its continuing intrinsic growth is paramount. There is work to do. The United States, for example, could be a much better leader in efforts to curb climate change. It's hard to fathom how Uncle Sam can keep a straight face while pressing countries such as China and India to reduce carbon emissions.[12] China and India have curtly rebuffed most such requests, and with due cause.[13] On a per capita basis, Americans are among the world's worst carbon emitters, at 18 metric tons of CO_2 per person per year, compared with China's 6.3 and India's 1.4.[14] It's laudable that China and India have begun to embrace environmental issues on their own, but America has to gain a much higher moral plateau before it can ask others to curb consumption. We should lead by example, whether it's to address climate change, global poverty, ethnic strife, or other global challenges.

Another way the developed world could be a model is through a more evolved set of aspirations. Imagine if greatness were rated not in terms of firepower or GDP but in terms of wisdom and intrinsic growth. We do this sometimes already. On issues such as gender equality, many developed countries are admirable, if still imperfect. Additional intrinsic growth would mean less material consumption and more involvement with self-transcendent ends. Other countries would likely follow.

Seeking our own growth also takes the edge off of paternalism. Humility is required in social causes, as privileged-world dogmas often cause damage. We should dispense with arrogant notions that we've reached some End of History.[15] Today's rich societies are, at best, adolescents with still a long way to go before they reach maturity. When everyone's intrinsic growth is a common goal, relationships become closer to true partnerships.

When Do We Intend to Start?

Isaac Asimov was tired of dark robot stories. Tales involving what he called the "Frankenstein syndrome" always had humanity destroyed by its own creations. So his novels mapped out an optimistic future where human beings prospered throughout the galaxy, often with the help of advanced technology.

Runaround, published in 1942, when Asimov was just twenty-two, is one example. It features a robot named Speedy who spends hours running in circles on Mercury, spouting nursery rhymes in an apparent malfunction. A human crew had dispatched Speedy after a fuel source, but the destination was contaminated with corrosive gas. Speedy thus keeps a fixed distance. It is the point of balance in Speedy's hardwired struggle between two conflicting commands: to retrieve fuel and to preserve itself. Realizing this, one of the crew hits on an idea. He opens up his spacesuit, letting in Mercury's harsh elements. Seeing the astronaut in danger causes Speedy to rush to the rescue, as his two competing orders are countermanded by a higher-priority imperative. Burnt into every robot brain in Asimov's universe is a rule that precedes all other commands, the First Law of Robotics: "A robot may not injure a human being, or, through inaction, allow a human being to come to harm."[16]

Speedy was an early model, but with it we already see the theme of technology-as-savior that permeates Asimov's stories. As Asimov grew up, so did his robots, but the savior complex remained. Each new story described a more sophisticated model, until many of the novels featured a godlike robot named R. Daneel Olivaw. Through self-repair, Daneel becomes immortal; and through future science, he learns extrasensory mind-control. Yet the First Law never leaves him. In the multi-millennial arc of Asimov's imagination, Daneel repeatedly rescues humanity from itself as it blunders toward an enduring galactic civilization.

There is a bit of Asimov in all of us. We want to believe that our technology, the fruit of our self-actualizing ingenuity, will save us from our own vices. The belief is both an acknowledgment that we need saving and a wish to be saved. Yet in clinging to this belief, we are renouncing our potential and our responsibility to save ourselves.

The flaw is not in either technology or technocracy, per se, but in our misguided, overly optimistic beliefs about what kinds of social change they will accomplish. It hasn't yet been a century since Asimov imagined his first fictional robots, but robots are already current news: Google has prototyped a self-driving car; software bots manipulate online product ratings; Amazon proposes delivery by automated quadcopter. These robots are designed for profit, not human betterment. Technology doesn't bootstrap an ethical outlook on its own. Ultimately, people govern technology. Any progress worthy of the name requires progress in human heart, mind, and will.

In spite of his optimism, Asimov–who served in the military during World War II and lived through the height of the Cold War–knew intimately that powerful technologies don't trump Stone Age emotions.[17] He worried that critics would see through his robot paternalism and pan him for painting human beings as a species in need of chaperoning. To this imagined criticism, he replied, "If we demand to be treated as adults, shouldn't we act like adults? And when do we intend to start?"[18]

ACKNOWLEDGMENTS

Like the ship of Theseus, the first draft of this book was replaced piece by piece, to the point that the final manuscript contains little of the initial text. I resisted many of the revisions, but looking back, I see how necessary they were, and for that I owe thanks to the many people who provided me with opportunity, advice, and critique.

AnnaLee Saxenian at the School of Information at the University of California, Berkeley, and Henry Tirri of Nokia Research gave me the priceless gift of one year's time and freedom to research and write the first draft. Its essence and argument have been preserved despite successive revisions.

Scott Stossel, magazine editor of *The Atlantic*, was generous to a fault in sharing feedback, expertise, contacts, and opportunities with me. This book would not have been published without him.

Several kind people in the publishing industry helped me along the way with no ultimate benefit to themselves. Among them, Howard Yoon and Melanie Tortoroli each gave precious input. A few agents and editors also offered thoughtful feedback, including Giles Anderson, Max Brockman, Joseph Calamia, Amy Caldwell, Laurie Harting, Jeff Kehoe, Rafe Sagalyn, Jeevan Sivasubramaniam, Anna Sproul-Latimer, Andrew Stuart, and Elizabeth Wales. Authors Ben Mezrich and Evgeny Morozov provided timely advice. Thank you.

I'm also grateful to Patrick Newell for inviting me to speak at the beautifully organized TEDxTokyo in 2010 (http://j.mp/ktTEDxTokyo).

The ideas in this book were fostered through close engagements with a number of organizations. I thank P. Anandan, Dan Ling, Rick Rashid, and Craig Mundie for the incredible opportunity to cofound Microsoft Research India, and to my colleagues in the Technology for Emerging Markets group for all of our research adventures – Ed Cutrell, Jonathan Donner, Rikin Gandhi, David Hutchful, Paul Javid, Indrani Medhi, Saurabh Panjwani, Udai Singh Pawar, Archana Prasad, Nimmi Rangaswamy, Aishwarya Lakshmi Ratan, Bill Thies, Rajesh Veeraraghavan, and Randy Wang. Teaching at Ashesi University was the chance of a lifetime, and it was Patrick Awuah, Nina Marini, and the Ashesi Class of 2005 who made that possible. The Venerable Tenzin Priyadarshi was kind to induct me as a fellow of the Dalai Lama Center for Ethics and Transformative Values at the Massachusetts Institute of Technology, and I have enjoyed speaking with other fellows. David Edelstein, Caroline Figueres, Rikin Gandhi, Bookda Gheisar, Dean Karlan, Deo Niyizonkiza, Bhagya Rangachar, and Cliff Schmidt each invited me to join their respective nonprofits – Grameen Foundation, IICD, Digital Green, Global Washington, Innovations for Poverty Action, Village Health Works, CLT, and Literacy Bridge – as board member or adviser, and I've learned a lot from an insider's view of their work.

Some sections rely on interviews with people who graciously shared their time: Regina Agyare, Patrick Awuah, Roy Baumeister, Abdul Mannan Choudhury, Jayshree Diggi, Ann Downer, Julia Driver, Esther Duflo, David Ellerman, Abraham George, Anirban Ghose, Chris Howard, Ron Inglehart, Deep Joshi, Neelima Khetan, Gary King, Kavitha L., Lalitha Law, Jorge Perez-Luna, A. V. M. Sahni, Barry Schwartz, Priyanka Singh, Tara Sreenivasa, Vera te Velde, Isaac Tuggun, and Mark Warschauer. I hope I've accurately represented their views, even where we might disagree.

The survey of Kenyan aspirations mentioned in Chapter 8 was conducted by Victor Rateng at Synovate. Thanks also to Shikoh Gitau and Joel Lehmann for help with the survey, and especially to Nathalia Rodriguez Vega for analysis.

Many people critiqued my drafts. A big "thank you" to all! Bill Thies and Suze Woolf went beyond anything I could have hoped for by providing meaningful notes on every chapter. Bill also helped me clarify a few tricky passages in detail. I'm also indebted to the following people for their in-depth critique of a few chapters: Nana Boateng, Henry Corrigan-Gibbs, James Davis, Mauricio Gonzalez De La Fuente, Ted McCarthy, Anita Prakash, Francisco Proenza, Roni Rosenfeld, and Eduardo Villanueva. Several professional editors – Simon Waxman, Jenna Free, Connie Chapman, and Christina Henry de Tessan – provided additional excellent suggestions.

I was blessed with many friends and acquaintances who each read one or more chapters: Shabnam Aggarwal, Varun Aggarwal, Jyotsna Agrawal, Michael Aldridge, Marika Arcese, Varun Arora, Sri Arumugam, Patrick Awuah, Sateesh Babu, Savita Bailur, Anton Bakalov, Rashmir Balasubramaniam, Eugene Bardach, Joanna Bargeron, Jason Belcher, Garima Bhatia, Lillian Bridges, Paolo Brunello, Fujin Butsudo, Cindy Chen, Gerry Chu, Melody Clark, Joshua Cohen, Carola Conces, Mo Corston-Oliver, Paul Currion, Melissa Densmore, Ron Dirkse, Jonathan Donner, Krittika D'Silva, George Durham, Hans-Juergen Engelbrecht, Lauri Ericson, Caroline Figueres, Sybille Fleischman, Rikin Gandhi, Ankur Garg, Maria Gargiulo, Anirban Ghose, Seshagiri GS, Leba Haber, Christopher Hoadley, Vigneswaran Ilavarasan, Ryan Jacobs, Susan Jeffords, Jofish Kaye, Itamar Kimchi, Anirudh Krishna,

Neha Kumar, Richa Kumar, Kimmo Kuusilinna, Susie J. Lee, Natalie Linnell, Andie Long, Tracey Lovejoy, Adnan Mahmud, Meghana Marathe, Derek Mathis, Indrani Medhi, Ghulam Murtaza, Satyajit Nath, Muchiri Nyaggah, Flavio Oliveira, Michael Paik, Diana Pallais, Saurabh Panjwani, Dan Perkel, Gretchen Philips, Sean Policarpio, Sammia Poveda, Abhishek Prateek, Barath Raghavan, Seema Ramchandani, Jon Rosenberg, Atsushi Sakahara, Sambit Satpathy, Jonathan Scanlon, Kevin Schofield, Frank Schott, Scott Stossel, Thomas Stossel, Jeff Swindle, Heather Thorne, Dan Toyama, Haruki Toyama, Toni Tsvetanova, Dipti Vaghela, Rama Vedashree, Rajesh Veeraraghavan, Jonathan Wai, Lowell Weiss, Renee Wittemeyer, Treena Wu, and Mel Young.

Additionally, my writing benefited from conversations and other forms of support from many, many people: Debbie Apsley, Özlem Ayduk, Marika Arcese, Siva Athreya, Garima Bhatia, Chris Blattman, Peter Blomquist, Maurizio Bricola, Jenna Burrell, Suvojit Chattopadhyay, Deepti Chittamuru, Magdalena Claro, Josh Cohen, Kristina Cordero, David Daballen, Kristen Dailey, John Danner, Ankhi Das, Alain de Janvry, Thad Dunning, Paolo Ficarelli, Greg Fischer, Bablu Ganguly, Maria Gargiulo, Achintya Ghosh, Rachel Glennerster, Richa Govil, Jürgen Hagmann, Naomi Handa-Williams, Saskia Harmsen, Gaël Hernández, Melissa Ho, Shanti Jayanthasri, Rob Jensen, Ashok Jhunjhunwala, Joseph Joy, Pritam Kabe, Ken Keniston, Neelima Khetan, Jessica Kiessel, Michael Kremer, Ramchandar Krishnamurthy, Antony Lekoitip, Miep Lenoir, Julia Lowe, Jeff MacKie-Mason, Drew McDermott, Patricia Mecheal, Pavithra Mehta, Ted Miguel, Eduardo Monge, Rohan Murty, Miguel Nussbaum, Chip Owen, Tapan Parikh, Paul Polak, Madhavi Raj, Ranjeet Ranade, Gautam Rao, Eric Ringger, Hans Rosling, Elisabeth Sadoulet, Maximiliano Santa Cruz Scantlebury, Jonathan Scanlon, Denise Senmartin, Jahanzeb Sherwani, Priyanka Singh, Pratima Stanton, Rick Szeliski, Steve Toben, Mike Trucano, Avinash Upadhyay, Dipti Vaghela, Suzanne van der Velden, Srikant Vasan, Wayan Vota, Terry Winograd, Christian Witt, Renee Wittemeyer, and Naa Lamle Wulff.

As much as I criticize technology hype, I'm not against technology per se. Let me own up to my big debts to digital technology: I wrote this book in Microsoft Word on an Asus laptop running Windows. My research was greatly facilitated by Google and Wikipedia. I found several out-of-print books on Amazon. Facebook allowed me to conduct informal surveys. And Twitter and other social media will play a part in book tours.

The title *Geek Heresy* comes from a May 6, 2001, article written about me by Tom Paulson, founder of the Humanosphere nonprofit news organization (see www.humanosphere.org). Tom has since become a close friend, and I've joined his nonprofit's board to support its uniquely fearless reporting about global development.

I'm immensely grateful to Clive Priddle at PublicAffairs and Jim Levine at Levine Greenberg Roston Literary Agency for seeing potential in my book and encouraging my aspirations. Through extensive revisions, Clive and his colleague Maria Goldverg made insightful recommendations and saved me from a glut of word count and self-indulgence. I'd also like to thank Melissa Raymond and Rachel King for production oversight, Kathy Streckfus for painstaking copyediting, and Pete Garceau and Cynthia Young for book design.

Any excesses that remain are due to my own stubbornness. I began this book with what I learned through research and personal experience, but as I pulled on the thread of the technological problems immediately in front of me, I found that it led me through a larger labyrinth that couldn't be understood in fragments. My earnest hope is that the whole, if fuzzy or flawed in the details, nevertheless presents an overall vision that is both coherent and compelling. Or at least thought-provoking.

Above all, I thank my wife, Jasmit Kaur, whose unfailing support – and willingness to read draft after draft after draft – brought out the best in me as I wrote this book.

APPENDIX: HIGHLIGHTED NONPROFITS

For readers who might be moved to support any of the inspiring nonprofits that I mentioned in this book, I provide the list below. All of them are exceptional in their respective areas, and each works to build heart, mind, and will. I receive no material compensation from any of them, but for full disclosure, an asterisk marks those on whose boards I sit. Of course, the list reflects my limited knowledge; it is not meant to exclude other worthy organizations.

- **Ashesi University** (www.ashesi.org) is a world-class, nonprofit, four-year university in Ghana focused on educating ethical, entrepreneurial African leaders.

- **Digital Green*** (www.digitalgreen.org) uses a unique video-based teaching methodology to improve agriculture, health, and nutrition in South Asia and Africa.

- **Innovations for Poverty Action*** (www.poverty-action.org) applies evidence gathered from randomized controlled trials to develop and scale up solutions for the developing world.

- **Pradan** (www.pradan.net) assists self-help groups to form, organize, and improve livelihoods of poor families in rural India.

- **Seva Mandir** (www.sevamandir.org) supports communities in southern Rajasthan in their effort to improve their lives via democratic, participatory development.

- **Shanti Bhavan** (www.shantibhavanonline.org) provides India's most disadvantaged children with a world-class education that emphasizes globally shared values and high career aspirations.

- **Technology Access Foundation** (www.techaccess.org) equips Washington State students of color for success in college and life through the power of a STEM education.

- **Village Health Works*** (www.villagehealthworks.org) delivers world-class, community-driven medical care and local development initiatives in the rural community of Kigutu, Burundi.

NOTES

Introduction

1. A full recording of the panel is available at Saxenian et al. (2011). The views that I presented there correspond roughly to Part 1.

2. "Tech-driven philanthropy" was the tagline on Google.org's home page (http://www.google.org) on the day of the panel, and as late as Nov. 14, 2012. It appears to have since changed, but when I Googled "tech-driven philanthropy" on Dec. 20, 2014, Google.org was still (mysteriously) the top hit.

3. International Telecommunications Union (2014); Ericsson (2014), p. 6.

4. World Wide Web Foundation (n.d.).

5. Page (2014).

6. Zuckerberg (2014). Internet.org's announcement is available at Internet .org (2013).

7. Duncan (2012).

8. Sachs (2008).

9. Clinton (2010).

10. DeNavas-Walt et al. (2009), p. 13, provide the US Census Bureau's graph of poverty. Incidentally, it seems that something quietly devastating began in the early 1970s. Commentators in a range of fields cite that period as the turning point where America (and possibly the Western world as a whole) began to decline. Hedrick Smith (2013) blames the 1971 Powell memorandum for turning corporations into narrowly selfish, power-hungry profit seekers. Political scientists Jacob Hacker and Paul Pierson (2010) blame a political system bent to the will of the wealthy. PayPal cofounder Peter Thiel (2012), 39:30, says technological advance has decelerated since the early 1970s (except in the computer industry). Economists Goldin and Katz (2009), p. 4, note that "educational advance slowed considerably for young adults beginning in the 1970s."

11. The evidence for middle-class income stagnation and rising inequality is well-established. See, for example, Piketty and Saez (2003) and US Department of Commerce, US Census Bureau (2011). Saez (2013) shows that the last time some

of these inequality measures were this high was in 1917. The facts are also largely uncontested–even fiscal conservatives such as Boudreaux and Perry (2013) agree with the statistics, even if they disagree about their causes and implications for policy.

12. CTIA (2011).

13. There is a chance that the poverty rate has been flat because nontechnological forces were increasing the rate of poverty from 1970 to now while technology was actually reducing it during the same period, and the two forces canceled each other out. If so, I'd be wrong that more technology by itself doesn't help social causes, but that would also mean that our social system tends toward greater poverty unless new technologies are invented at a breakneck pace. That is an even darker scenario, which, if true, would only further justify the overall thesis of Part 2: that we need to pay more attention to social forces rather than to technological ones.

14. Carolina for Kibera (n.d.).

15. Of course, it's understandable that corporate spin highlights products even if executives praise employee talent. The problem occurs when the rest of society drinks the Kool-Aid. And it does. I once had a conversation with an influential Harvard development economist in which I mentioned the importance of growing wisdom in people. He fixed me with a quizzical look and asked, "How is that different from what you'd want for your kids?" He seemed to believe that what was good for international development ought to be fundamentally different from what was good for his family.

16. Denshi Burokku ("electronic block" in Japanese) was an educational toy popular in Japan during the 1970s and 1980s. They were discontinued in 1986, but have since been periodically reissued.

17. Viola and Jones (2001).

18. Criminisi et al. (2004).

19. Rowan (2010) tells the story behind the Kinect system; Toyama and Blake (2001) describe the technology.

20. Microsoft has a larger research lab in China, but it is based in Beijing–which, with its gleaming skyscrapers and slum-free environs, is difficult to classify as "developing world." In contrast, unattended cows regularly walked by our center in India, and the neighborhood saw plenty of tarpaulin tents housing migrant workers.

21. The brilliantly chosen WEIRD acronym was introduced by Henrich et al. (2010), who argue that most psychology studies are conducted on rich-world undergrads, an unrepresentative slice of the global human population.

22. Plato (1956), pp. 64–65, refers to the self-animated "statues of Daidolos," which "must be fastened up, if you want to keep them; or else they are off and away."

Chapter 1: No Laptop Left Behind

Conflicting Results in Educational Technology

1. Pal et al. (2006). Joyojeet Pal visited eighteen schools in four states of India, with help from the Azim Premji Foundation. Pal (2005) maintains photographs and a slide presentation about his visits.

2. Pawar et al. (2007). MultiPoint was one of the first projects at Microsoft Research India that went through the full cycle of our approach to research: immersion in a specific environment; iterated prototyping and exploratory field trials; confirmatory evaluation; and ultimately, technology transfer and productization.

3. United Nations (2005).

4. Negroponte frequently repeats this mantra in public appearances. It also appears on the "Mission" page of the One Laptop Per Child (n.d.) website.

5. Surana et al. (2008) measured power surges as high as 1,000 volts in the rural Indian power grid; most consumer electronics are not rated above 240 volts.

6. The projects mentioned in this paragraph are a subset of the education-related projects that researchers in my group at Microsoft Research India conducted. The projects varied in their outcomes, but every single one of them saw something of the Law of Amplification to be described in Chapter 2, namely, that the pedagogical capacity of the school and teachers were critical to the technology having an impact. The following references match the order of their mention in this paragraph: Sahni et al. (2008); Paruthi and Thies (2011); Panjwani et al. (2010); Hutchful et al. (2010); Linnell et al. (2011); Kumar (2008).

7. Cuban (1986) provides a thorough deconstruction of the history of electronic technologies in America. The quotation by Edison appears in Weir (1922).

8. Darrow (1932), p. 79.

9. Oppenheimer (2003), p. 5.

10. Santiago et al. (2010); Cristia et al. (2012). The studies found no increase in either mathematical or verbal academic achievement, but they did find that cognitive skills as measured by Raven's Progressive Matrices, a test of spatial-visual ability, did increase significantly. The study demonstrates Chapter 2's Law of Amplification exactly: Children have a natural curiosity and desire to grow their cognitive skills through play, and computers can amplify that. However, the directed motivation required for educational achievement requires strong pedagogy before technology can help.

11. De Melo et al. (2014), in Spanish. An English overview of the results with commentary appears in Murphy (2014b).

12. Linden (2008); Barrera-Osorio and Linden (2009). Linden's studies are among the first to apply large-scale randomized controlled trials to measure the impact of computers in developing-world schools.

13. Behar (2010). Behar continues to be an outspoken critic of "silver bullets in education"; see Behar (2012).

14. The examples and quotations in this section are taken from Warschauer et al. (2004).

15. To be fair to Warschauer, I should mention that in private communication, he indicated some discomfort with my characterization of his work. If I understand him correctly, it's not that I am misrepresenting any particular point he makes, but that my overall presentation fails to emphasize that computers can have positive impacts in well-run schools. I have tried to present a balanced perspective in this chapter, and I am not quoting Warschauer out of context. Any residual misrepresentations of his view are my fault. As to the underlying thesis that technology amplifies institutional capacity–a point described in greater depth in Chapter 2 and which concisely explains both technology's potential and its failures–Warschauer and I seem to agree.

16. Bauerlein (2009), p. 139.

17. Oppenheimer (2003). *The Flickering Mind* continues to be among the best critiques ever written about computing technology in education.

18. Hu (2007).

19. Duncan (2012).

20. Prensky (2011), p. 9. I don't deny the possibility that video games can be used for productive educational purposes, and educational video games are worth exploring further. But the evidence for their value is scant. And on top of that, it's not clear that even if the short-term learning were effective, we'd want to raise a generation of people who can only learn if material is presented as a video game. Part of the point of a deep education is to learn how to learn–even when the material is *not* engaging–or, alternatively, to learn how to make otherwise dull material interesting for yourself. You can't learn these skills if every learning opportunity is entertaining.

21. Wood (2013).

22. Sanders (2013).

23. Mitra and Dangwal (2010).

24. Warschauer (2003) and Arora (2010) base their skepticism of the Hole-in-the-Wall program on personal visits. Meanwhile, the only studies that show positive impact of Hole-in-the-Wall installations are methodologically questionable ones conducted by Mitra and his colleagues.

25. Mitra and Arora (2010).

26. Fairlie and Robinson's (2013) study is among the few randomized controlled trials of the educational effect of a personal laptop for students, and the results are definitive. Two of the organizations that funded the study–Computers for Classrooms and ZeroDivide Foundation–are nonprofits whose mission is to

increase home computers for families without them, so they had undoubtedly hoped for a different outcome. Their willingness to fund such research and have the results published is admirable, and it gives additional credibility to the results.

27. Duncan (2012).

28. Ibid.

29. The Organisation for Economic Co-operation and Development's (OECD's) (2010b) summary of what makes schools successful is striking in its lack of mention of computers or other technology. The results for China are actually results just for the city of Shanghai. As of 2012, PISA tests have not yet been administered to China as a whole country.

30. OECD (2010b), p. 106.

31. Some of the paragraphs in this chapter are either verbatim excerpts or adapted sections of Toyama (2011).

32. CBS News (2007). Negroponte, however, conducted no serious study of educational gains. His excitement was based on such things as the fact that families used the laptop as a source of light in the evenings, and that their first English word was supposedly "Google."

33. Warschauer (2006), pp. 62–83.

34. Sinclair (1934 [1994]), p. 109.

Chapter 2: The Law of Amplification

A Simple But Powerful Theory of Technology's Social Impact

1. Rangaswamy (2009).

2. Heilbroner's (1967) article is among the most cited in the technology and society literature, probably because it is one of the few in which a respected scholar sides unabashedly with technological determinism. Heilbroner (1994) later softened his stance, but only slightly. Today, few scholars admit to pure technological determinism, but as the examples later in this chapter show, there are plenty of influential nonacademics who subscribe to its views. MacKenzie and Wajcman (1985) refer to technological determinism as the "single most influential theory of the relationship between technology and society."

3. Feenberg (1999), p. 78.

4. *Star Trek: First Contact* (1996).

5. Schmidt and Cohen (2013), p. 257. Their book takes pains to concede the dark side of technology, but the concessions are raised only to disarm the reader into accepting their larger thesis: More technology is better. Toyama (2013a) reviews the book in more detail.

6. Shirky (2010).

7. These quotations are from *Economist* (2008) and Diamandis and Kotler (2012), p. 6.

8. In a survey by the US Department of Agriculture, Coleman-Jensen et al. (2013) show that in 2012, 7 million households in America had "very low food security," with, for example, 97 percent of those households reporting that "the food they bought just did not last and they did not have money to get more." Those households included 4.8 million children.

9. Food and Agriculture Organization (2013).

10. Morozov (2011), p. 88. Morozov provides a much-needed overview of the dark side of the Internet in repressive politics around the world.

11. Ibid., p. 146. The original quotation appears in Dahl (2010).

12. Ellul (1965 [1973]), p. 87.

13. Postman (1985 [2005]) makes a strong case that as a result of television, modern society has begun to judge everything by its entertainment value. His analysis is astute, and it applies even more in the Internet age. Postman tends toward a kind of technological determinism, however, and blames the technology itself for societal trends. I would argue that an inclination for entertainment exists within us, and that technology's role is to amplify it. The question is whether we as a society could amplify other aspects of ourselves without turning into YouTube-obsessed vegetables.

14. Jasanoff (2002).

15. Malmodin et al. (2010).

16. Delforge (2014) estimates data-center electricity use in 2013 at 91 billion kilowatt-hours–about 2.25 percent of total US electricity usage, which hovers around 4,000 billion kilowatt-hours (US Energy Information Administration 2014b). This closely tracks with Koomey (2011), who put the figure at 2 percent in 2011.

17. Skeptics have derided utopians as starry-eyed, and in response, utopians have learned to tone down their rhetoric–but not entirely convincingly. Sometimes, the attempts are clumsy, as when Schmidt and Cohen (2013), p. 257, write, "The case for optimism lies not in sci-fi gadgets or holograms, but in the check that technology and connectivity bring against the abuses, suffering and destruction in our world." Translation: The case for optimism isn't in technology, but it's in technology. Others are more careful. I scoured two of Shirky's (2010, 2011) books and found no instances in which he credits social benefit to any technology outright. Yet it's undeniable that he's gaga for technology. The conceit behind his book *Cognitive Surplus* is that, thanks to the participatory nature of the Internet, couch potatoes worldwide will stop wasting billions of hours watching TV–that's the surplus of the title–and instead put some of that time to good use.

18. Linda Stone (2008) popularized the notion that we maintain "continuous partial attention" with our many digital technologies. She did not necessarily mean it in the derogatory way I suggest here.

19. Jensen (1998 [2005]), p. 252.

20. Carr (2011), p. 224.

21. Ellul (1964), p. xxxi.

22. Kranzberg (1986).

23. Latour (1991). In the academic field called "science and technology studies," there's a cottage industry stating and restating various versions of contextualism. These kinds of theories are occasionally profound, but often they're just unedifying. Among the field's most popular ideas is one promoted–and sometimes self-criticized–by influential French sociologist Bruno Latour. He helped develop a concept called Actor-Network Theory, in which people and technologies are nodes that affect one another in a fluid web of interconnected relationships. Latour (1991) describes it like this: "If we display a socio-technical network–defining trajectories by actants' association and substitution, defining actants by all the trajectories in which they enter, by following translations and, finally, by varying the observer's point of view–we have no need to look for any additional causes. The explanation emerges once the description is saturated." Even for describing something as simple as a hotel key chain, though, the networks quickly become a tangled Gordian Knot, and Latour insists that the only way to understand the whole is by carefully tracing every last strand. So, on the one hand, you have a richer description. On the other hand, that's all you have. Concise explanation or understanding is not forthcoming.

24. Veeraraghavan et al. (2009).

25. In a paper we titled "Where There's a Will, There's a Way," Smyth et al. (2010) found that many low-literate young urban adults in India are facile users of Bluetooth file exchange on their phones despite the English interfaces and complex steps required. They were driven by a strong desire to trade music and movie files. Poor user interfaces were not an obstacle to usage.

26. The phrase "social determinism" as used in this book always refers to the idea that technology's impact is determined by human forces. It should not be confused with another definition of social determinism, in which individual human behavior is believed to be caused entirely by social and cultural forces, and not by physical or biological ones.

27. Autonomous robots–physical or virtual–could be said to act on their own, but even then, it will be people who designed and directed the robot personalities, or the processes by which they think. Those robots may end up acting in a way we didn't wholly intend, of course. In Chapter 3, I'll address the nature of unintended consequences.

28. Medhi et al. (2007).

29. Medhi et al. (2013). The description of the experiment is simplified in these paragraphs: The actual experiment involved three different interfaces, involving two nested interfaces of different depths.

30. The studies referred to are, respectively, Findlater et al. (2009), Chew et al. (2011), and DeRenzi et al. (2012).

31. Amplification is notably absent in the field of science and technology studies; scholars in that field tend to have a low opinion of instrumental theories of technology. In the information systems literature, one theory that comes close to amplification is absorptive capacity theory, as first articulated by Cohen and Levinthal (1990), which argues that an organization's ability to absorb technology determines what it can do with it.

32. I first wrote about amplification as it applies to technology and poverty in Toyama (2010).

33. Linden (2008); Santiago et al. (2010).

34. Some people emphasize the value of play in childrearing, and play is certainly important. Some amount of video games and social media may serve that purpose and nurture children in some intangible ways. But digital recreation is not sufficient for anything like a true K-12 education any more than extended time at a jungle gym in and of itself produces Olympic athletes.

35. Rao (2011).

36. Hauslohner (2011).

37. Cohen (2011).

38. Tsotsis (2011).

39. CNN (2011).

40. Olivarez-Giles (2011).

41. Kirkpatrick and El-Naggar (2011).

42. Chulov (2012).

43. Anyone interested in the real forces of revolution and suppression in the Middle East would profit by reading Madawi Al-Rasheed's work. She strikes the right balance in acknowledging the role of social media while consistently returning to essential political and cultural forces as the underlying explanations for high-visibility events and non-events. This quotation and the descriptions in the previous paragraph were taken from Al-Rasheed (2012).

44. Lee and Weinthal (2011).

45. Paul Revere Heritage Project (n.d.). In highlighting the lanterns, Henry Wadsworth Longfellow, whose "Midnight Ride" is the basis for this American legend, was applying poetic license. After all, there's romance in ingenuity; the symbolism of lamplight breaks up the monotony of what would otherwise have just been one long ride on a horse. But while creative license leads to catchy ideas that serve poetry, it's not a good basis for accurate analysis or sound policy.

46. *60 Minutes* (2011).

47. Morozov (2011).

48. Clay Shirky makes these exuberant comments in an interview with TED owner and curator Chris Anderson (2009).

49. Gladwell (2011).

50. Taylor (2011).

51. Yaqoob and Collins (2011).

52. Tichenor et al. (1970).

53. Mumford (1966), p. 9.

54. Agre (2002). Agre appears to be the first person to have outlined a theory of amplification for technology and society. I agree with almost everything he has written on the topic, so we differ only on emphasis: First, Agre claims that the holistic effect of the Internet on politics is impossible to predict, because the underlying forces are so complex; I agree but believe that prediction is possible in more limited cases where human forces are easier to understand. Second, Agre confines himself to a discussion of the Internet in politics and governance; I believe amplification applies to all of society's interactions with a broad range of technologies, not just the Internet, and not even just digital.

Chapter 3: Geek Myths Debunked

Dispelling Misguided Beliefs About Technology

1. Harvey (1988), for example, discusses the advertising strategy for marketing the Walkman to children. In it, he notes, "My First Sony [Walkman] has created a new merchandise category for toy stores," and, "The company [Sony] has a long history of pushing through products it believes in" over the doubts of distributors. Remarks like these are readily taken up as proof of a technology firm's ability to arbitrarily alter consumer behavior, as noted in Sanderson and Uzumeri (1995): "It is not uncommon to view innovative success as the natural result of managerial leadership and effective marketing."

2. For a thorough treatment of the cultural studies angle to the Walkman, see Du Gay (1997). The reproduced readings at the end of Du Gay's book show the range of approaches to the Walkman, most of which, incidentally, are not inconsistent with the Law of Amplification.

3. In fact, there are niche firms that sell hairshirts and such (search online for "cilice")–illustrating the rich variety of human culture–but they are hardly mainstream.

4. Turkle (2011).

5. See, for example, Baym (2010), pp. 51–57.

6. Rosenfeld and Thomas (2012).

7. Wortham (2011); Leland (2011). Przybylski et al. (2013) reveal that those with lower life satisfaction and satisfaction of social needs tend to display more FOMO-related behaviors on social media.

8. Goldman (2009).

9. GOP Doctors Caucus (n.d.).

10. Reinhardt (2012).

11. White (2007); Rumpf et al. (2011).

12. Organisation for Economic Co-operation and Development (2013a). For an insightful analysis of American health care, including comparisons with other countries' systems, see Cohn (2008).

13. See, for example, Spiceworks (2014).

14. Brill (2013).

15. Hampton et al. (2011).

16. See, for example, Crescenzi et al. (2013), Lee et al. (2010), and Olson and Olson (2000). The last provides excellent summary and analysis for why distance doesn't collapse even with digital technologies.

17. Cairncross (1997), p. xvi. Presumably in a response to critics, Cairncross (2011) softens her points in a revised edition. The corresponding sentence becomes: "People will communicate more freely with human beings on other parts of the globe. As a result, while wars will still be fought, the effect *may* be to foster world peace" (emphasis mine, recall tactics by Schmidt and Cohen 2013). But her general thrust remains much the same–in fact, she adds a few more ways in which technology will definitely improve the world, such as in the developing world.

18. Van Alstyne and Brynjolfsson (2005).

19. Selective exposure goes back to work by seminal psychologist Leon Festinger (1957), who posited the idea of cognitive dissonance–the discomfort people feel when presented with contradictory information. Selective exposure occurs when, in a bid to avoid cognitive dissonance, people try to limit their exposure to information that confirms their beliefs.

20. Van Alstyne and Brynjolfsson (2005).

21. Stecklow (2005).

22. Mukul (2006); Raina and Timmons (2011).

23. A phablet is bigger than a smartphone, but smaller than a tablet.

24. That the digital divide is a symptom of other socioeconomic divides was astutely noted about telecenters by *Economist* (2005). The same article, however, curiously went on to suggest that mobile phones would somehow "promote bottom-up development" that presumably would help close socioeconomic divides because of their greater penetration. In other words, the telecenter-based digital divide is a symptom of socioeconomic divides, but the mobile-phone-based digital divide is not.

25. This paragraph argues that the absolute difference in outcomes between high-and low-capacity people increases with an even spread of technology. The *relative* difference may not change. But! If you fold in the fact that richer people access superior technologies, then the increase in inequality is superlinear: As new technologies appear, the rich get richer out of proportion to their initial relative wealth. I don't mean to say that low-cost technologies can't help poorer people–they certainly can. And for some people, like the political philosopher John Rawls, this would be good enough, at least in theory. In practice, though, this argument neglects the fact that political power and finite natural resources are both zero sum–the more that someone has, the less others do, so increasing absolute inequality is *necessarily* worse for those at the bottom. In any case, for anyone who sees inequality itself as the problem, low-cost technology is in no way the solution.

26. Much of the text that follows about Gary King's studies previously appeared in Toyama (2013b).

27. King et al. (2013a).

28. King et al. (2013b).

29. King et al. (2013a) cite an example of apolitical collective action: In 2011, following Japan's nuclear plant disaster, there was a rumor that iodized salt protected against radiation. Online posts that might incite hordes of shoppers to buy salt were suppressed.

30. The quotation is from King (2013a), summarizing Dimitrov (2008).

31. Guilford (2013).

32. Brewis et al. (2011) suggest that there is a global trend toward stigmatizing obesity, but it overviews a range of work showing that different cultures have different weight preferences. Sobal and Stunkard (1989) review literature linking socioeconomic class to weight.

33. Lenhart (2012); Hafner (2009). In 2011, when I gave a talk at the Maricopa Community Colleges, I asked how many people sent more than one hundred texts a day. Almost every student raised a hand while the faculty looked around in disbelief.

34. Though, with amplification in mind, this still seems like a predictable case of amplified teenage socializing.

35. Thanks to skeptics, it's clear that just about every technology has outcomes not wholly intended by its creators or its users. To put a new technology out there *is* to cause outcomes no one entity can wholly predict. Thus, routine failures to anticipate them, monitor them, and manage them are again a kind of passive intention.

36. Sartre (1957 [1983]), p. 15: "Man is nothing else but what he makes of himself."

Chapter 4: Shrink-Wrapped Quick Fixes

Technology as an Exemplar of the Packaged Intervention

1. My use of the term "packaged interventions" is very similar to what Evgeny Morozov (2013) calls "technological solutions" and what William Easterly (2014) calls "technical" or "technocratic solutions." I didn't want to use the word "solution," though, since packaged interventions are *not* necessarily solutions. And I avoid "technology" and "technocracy" because I already use them elsewhere to mean specific things.

2. Yunus (1999), p. 48.

3. From Yunus's Foreword in Counts (2008), p. viii.

4. Counts (2008), p. 4.

5. Based on data available at MixMarket (2014).

6. Bloomberg Businessweek (2007).

7. In 2010, a microlending organization called SKS Microfinance repeated Compartamos's feat in India, raising $358 million in its IPO, and setting off a national debate that crippled the microcredit industry in that country. Many accused SKS and other microfinance institutions of pushing loans too aggressively. Vijay Mahajan, an elder statesman of microfinance in India, said that some of the newer organizations "kept piling on more loans in the same geographies. . . . That led to more indebtedness, and in some cases it led to suicides" (Polgreen and Bajaj 2010). Angry politicians in the state of Andhra Pradesh passed strict laws on how microloans can be issued and instigated a grassroots backlash. Borrowers stopped repayments altogether. See also Bajaj (2011). An attempt at a levelheaded assessment of these situations is offered in Rosenberg (2007).

8. Yunus (2011).

9. Collins et al. (2009).

10. David Roodman (2012) at the Center for Global Development performs an excellent deconstruction of microfinance, incorporating and analyzing recent studies without oversimplification. Many of the studies cited in this chapter are given more detailed treatment in his book, which also explains what is firmly established about microfinance and what remains unknown.

11. Karlan and Zinman (2010).

12. Karlan and Zinman (2011).

13. Angelucci et al. (2013).

14. Banerjee et al. (2010). In 2005, economists Abhijit Banerjee, Esther Duflo, Rachel Glennerster, and Cynthia Kinnan persuaded a microlending organization called Spandana to open branches in 52 locations, randomly selected from a larger pool of 104 poor, urban neighborhoods in Hyderabad. Fifteen to 18 months after the branch openings, the neighborhoods where microcredit was available didn't appear on the whole to be wealthier than those without. Nor were there detectable

changes in overall household spending, women's say in spending decisions, health-related measures, or educational outcomes. The most they could say was that Spandana's presence caused a 2 percent increase in the number of households that opened new businesses, a 55-cent increase in spending on durable goods per person per month, and a 23-cent decrease in perishable consumer goods–durable goods being more likely to support businesses. Their conclusion? "Microcredit therefore may not be the miracle that is sometimes claimed on its behalf, but it does allow households to borrow, invest, and create and expand businesses."

15. Ray and Ghahremani (2014).

16. Vornovytsky et al. (2011).

17. See, for example, Krugman (2010).

18. Banerjee et al. (2010); Drexler et al. (2010); Karlan and Zinman (2011).

19. This generalized conception of technology is not rare. Economists, for example, routinely speak of structures of human organization as a kind of technology. However, since most people think of technology as being a physical artifact, I use the phrase "packaged intervention" as its generalization.

20. In case you believe these are rare or short-term effects, consider the different trajectories of post-Soviet countries. Lithuania, Latvia, Estonia, and possibly Ukraine have made a lasting transition to democracy, but the rest have backslid into dictatorships, both real and virtual, despite initial elections. Even in democracies with peaceful, elected turnovers of power, long-held attitudes can impede governance. When I was in India, I found its claim to be the world's largest democracy to be a bit of an exaggeration. What I saw was more like a feudal system with term limits. Politicians act like dukes and barons when in power, and most citizens happily oblige, bowing low to officeholders and paying *baksheesh* for routine government services. Everyone knows that government employees have an informal income stream, but by many it's accepted as a privilege of power–even one to aspire to. And everyone knows that everyone knows. In 2010, when B. S. Yeddyurappa, then the chief minister of Karnataka, was taking heat for excessive government corruption, he reprimanded his own administration with a wink *on the public record*: "Let all of us stop making money for ourselves. All of us should now work for Karnataka" (IBNLive 2010).

21. Seymour Lipset (1959, 1960) was among the first to argue that a range of socioeconomic properties appear to encourage democracy, an argument to which I will return in Chapter 9 in a modified form. Political scientist Robert Dahl (1971) focuses on eight institutional requirements for democracy, among which are political parties, the right to run for office, a free press, associational autonomy, the rule of law, and an efficient bureaucracy.

22. See, for example, Achebe's (1977) takedown of Joseph Conrad's *Heart of Darkness*.

23. Achebe (2011).

24. *Atlantic* (2012).

25. Porter (2013) reports that women of prime working age earn only about 80 percent of what their male peers earn.

26. The laptop-as-vaccine statement was made by Negroponte (2008) at a TED talk about One Laptop Per Child. He repeated the same claim when he and I were on a panel at MIT (*Boston Review* 2010). He must have felt that the analogy resonated.

27. From the Global Polio Eradication Annual Report (World Health Organization 2011). It's understandable that polio eradication efforts go poorly in areas with open conflict, such as Afghanistan or Nigeria. But even where violence is less frequent, as in western Chad, the report notes, "operational issues are the main reason children are still being missed by vaccination campaigns, although social and communication problems are also important, particularly in key high-risk areas."

28. It is generally agreed that smallpox is easier to eradicate than polio because smallpox always results in visible symptoms. With polio, for every person who shows symptoms, hundreds of others carry and spread it without symptoms. Thus, the only sure way to eradicate polio in a given region is to vaccinate everyone in it. How feasible that is depends not on technology, but on the reach and quality of government administration and health-care institutions.

29. Yunus (2011).

30. Bloomberg Businessweek (2007).

31. Yunus (1999), p. 205.

32. Ibid., p. 140.

33. Consultative Group to Assist the Poor (2008).

34. Banerjee et al. (2010).

35. Tripathi (2006).

36. Fears of sterilization occur regularly in the developing world, probably based on a past history of forced vasectomies in some countries (Population Research Institute 1998). Similar fears resurfaced recently in Pakistan (e.g., Khan 2013) and in Kenya (e.g., Gander 2014). As for vaccine fears in the developed world, see, for example, Mnookin (2011).

37. It can sometimes be helpful to project what would happen with technologies of the future. Put aside your conscience, and imagine a future in which a technology called "the Avatar" becomes available. It is a special chip that can be surgically implanted into a person's spinal cord just below the cerebellum, enabling the total hijacking of a person's voluntary muscle system via wireless command. (The spine's information bandwidth is probably around 16 megabits per second, only slightly more than is promised by current 3G networks.) Health-care workers could be

turned into remote-control puppets who visit assigned households faithfully, execute engagement scripts flawlessly, and never give in to laziness or corruption. In other words, this is a technology that would overcome all the pesky problems of messy human behavior. Yet, as powerful a technology as the Avatar might be, even *it* would require a good system for delivery and implementation. Not only would the technology have to be surgically installed on a person-by-person basis, but someone would have to maintain the technology, restock broken units, monitor performance, and deal with the predictable problem of subversive hosts seeking to rid themselves of the devices. And each of these activities would require solid implementation. In short, there would have to be ongoing institutional support beyond the packaged intervention itself, and at a much higher price than the cost of hardware. Of course, this dark potential future is only satire, but the point is that even an absolute technology requires strong human implementation to work. Similar lessons naturally hold for far less powerful packaged interventions.

38. American Sociological Association (2006).

39. Rossi (1987).

40. There is an implicit fourth problem that Rossi mentions but resigns himself to. It corresponds to what's required of beneficiaries of packaged interventions. Rossi recognized that no program works without motivation and capacity among those whom it is meant to help, but he felt it was beyond intentional policy to address this deficiency, saying, "It is likely that large scale personality changes are beyond the reach of social policy institutions in a democratic society." This is a critical point that is addressed in Part 2–I believe he gave up too easily.

41. Yunus (1999), p. 140.

42. Ibid., p. 205.

43. Wikipedia (n.d.), "FINCA International," http://en.wikipedia.org/wiki/FINCA_International.

44. Kiva.org (n.d.).

45. Opportunity International (n.d.).

46. Yunus (1999), pp. 135–137.

47. Based on data available at MixMarket (2014). The estimate is low because it includes only organizations registered with the exchange at the time and excludes microcredit activities in the developed world.

48. Heeks (2009).

49. Organisation for Economic Co-operation and Development (OECD) (2014a). The figure cited includes all bilateral aid from OECD countries and multilateral aid from organizations like UNICEF.

50. See Fraser (2012) and Plumer (2012) for critical reporting on cookstoves.

51. The performance of charter schools is an ongoing debate, but it seems fair to say that the results are mixed. A good summary, with references to primary

research, is offered by Ravitch (2011), pp. 138–143. Ravitch's book, incidentally, offers a brilliant argument against packaged interventions in American education, based on her life's work as an educator and education researcher.

Chapter 5: Technocratic Orthodoxy

The Pervasive Biases of Modern Do-Gooding

1. George Packer (2014) cites an estimate of $5.25 billion in book sales for Amazon in 2013, and Milliot (2014) estimates total book sales at $15 billion. Packer also notes that in 2010, Amazon captured 90 percent of e-book sales.

2. The reference to Orwell was pointed out by Streitfeld (2014). The quotation is from Orwell (1936).

3. Barnes & Noble Booksellers (n.d.).

4. Thompson (2010) offers a rich history and analysis of the book industry in the four decades since about 1970.

5. Thompson (2010), pp. 389–392, describes this trend in detail and calls it a "winner-takes-more market."

6. In respective, eponymous books, Chris Anderson (2008) describes the long tail, and Robert Frank and Philip Cook (1996), the winner-take-all society.

7. Duflo et al. (2012).

8. In another paper with different colleagues, Duflo herself writes, "[The] recent evidence suggests that many interventions which increase school participation do not improve test scores for the average student. Students often seem not to learn anything in the additional days that they spend at school" (Banerjee et al. 2007). As the next few paragraphs explain, if you accept the generalized conclusions implied by the two papers, there is a contradiction that isn't addressed by either one.

9. The authors note in a footnote that tamper-proofing only meant placing "heavy tape" over the controls.

10. Duflo et al. (2012). "Nonformal education" is what Seva Mandir calls its program, in contrast with the formal government education system for which the children are being prepared. Despite the name, the pedagogy is formal and modeled on good classroom teaching.

11. Abdul Latif Jameel Poverty Action Lab (n.d.).

12. In addition, many of the best RCTs are overseen by high-caliber researchers like Duflo–yet another unusual circumstance that doesn't come with the packaged intervention itself.

13. For excellent tips on good classroom interaction, see Doug Lemov's (2010) *Teach Like a Champion*. I have found his book to be invaluable for teaching K-12.

14. When an RCT carried out in partnership with Organization X results in data suggesting that Program Y has an impact, what it proves is not that Y has

impact on its own, but that Y has impact if carried out by an organization like X. All too often, X and Y are *both* necessary for impact, just as Seva Mandir's efforts and the camera monitoring were both needed for impact. In their conclusion, Duflo et al. (2012) consider whether their results would generalize to other schools, and they optimistically suggest that it would. They write, "Our results suggest that providing incentives for attendance in nonformal schools can increase learning levels." There is just one nod to external vailidity in the paper. They mention, "The question arises, however, as to whether incentive programs can be instituted for government teachers, who tend to be politically powerful. It may prove difficult to institute a system in which they would be monitored daily using a camera or similar device. Our findings suggest, however, that the barriers currently preventing teachers from attending school regularly (e.g., distance, other activities) are not insurmountable. Given political will, it is possible that solutions to the absence problem could be found in government schools as well." The wording is careful not to overreach, yet in the way of technological optimists, they hide any reservations behind the word "can," as noted by Toyama (2013a).

I exchanged emails with Duflo on June 24–25, 2014, to seek her response. In the hyper-confident tone characteristic of many economists, she wrote, "I absolutely stand by the conclusion of the paper, and I don't really see why the fact that Seva Mandir was trying to improve (all) their schools in other ways as well makes it less externally valid than if they were not." If Seva Mandir schools differ from other schools, obviously, it has an impact on external validity. Our differences, then, come down to whether Seva Mandir's teaching and management are atypical. I thought they were exceptional in comparison to many rural government schools. Duflo dismisses the possibility: "Seva Mandir teaching is not so spectacular. It was frankly bad before they stepped up a bit."

My larger critique with RCTs is not with the methodology itself, but with the tendency of experimenters to generalize too confidently and to be too cavalier with external validity.

15. Banerjee and Duflo (2011), many places, but, for example, p. 272.

16. A Georgetown economics professor and former director of the World Bank's Development Research Group, Martin Ravallion (2011), best summarizes some of the difficulties with RCTs. A more academic critique is offered by Nobel economist James Heckman (1997). The problem I raise in this section is a special instance of the external validity problem as well as the tendency to do experiments that are convenient to run, but not necessarily the most revealing.

17. Most RCTs focus on evaluating the effectiveness of specific packaged interventions. I'm not aware of any RCTs to date that test the idea that a capable organization is the key to making a given packaged intervention work. Such a study could be done, in theory, but to remain within the epistemological constraints of hard-core

randomistas, it would require a treatment that measurably improved organizational capability within the duration of the study. An experiment like that would not be cheap or easy to implement, especially because organizational capability takes a long time to improve and is difficult to measure. These are severe practical limitations of RCTs. The studies that get done are invariably the ones that are cheaper, shorter term, easier to find metrics for, and easier to run.

18. Rossi (1987).

19. If I could change one thing about RCTs conducted by economists, it would be this: The larger context in which the RCT was run should be reported in detail, and there should be explicit, thorough discussion about expected external validity. What are the relevant aspects of local culture, history, geography, climate, etc.? If there were partner organizations involved in the trial, what were their unique strengths and weaknesses when compared with similar organizations? Under what conditions should readers expect to see similar outcomes?

20. Prahalad (2004).

21. Ibid., p. 16.

22. Ibid., pp. 4–16.

23. Ibid., p. 4.

24. Karnani (2007), p. 93, Table 1.

25. It also turns out that low-cost sachets are not great for business after all. Since Prahalad's book, soap, shampoo, and all manner of detergent were increasingly sold in small, low-cost packets. This set off multiple price wars between HLL and Proctor & Gamble, resulting in the odd phenomenon that the price per volume of shampoo, for example, became lower in small sachets than in large bottles. An HLL executive once confessed to me that they gained market share but lost in terms of absolute net profits; they wanted to get out of the sachet business but couldn't see a way to do so. Indeed, a search online for "India sachet price war" returns many articles suggesting this point.

26. Ibid.

27. Yunus (2007).

28. Toms (n.d.). The exact wording on the company website as of September 2014 was "One for One: With every product you purchase, Toms will help a person in need."

29. Toms (n.d.). The one-for-one model means they must have sold 10 million pairs to donate as many. Their shoes are priced between $40 and $100, which translates to $400 million to $1 billion in total revenue.

30. Bansal (2012) and Butler (2014) provide good summaries of Toms Shoes criticism. Wydick et al. (2014) ran an RCT to test whether shoe donations caused households to buy fewer shoes, but with inconclusive results. Murphy (2014a) interprets the results in context of the criticism.

31. Toms (n.d.). Toms began with manufacturing in China, and has since expanded to Ethiopia and Argentina. Mycoskie has also announced a plan to start a factory in Haiti. In more than thirty other countries where Toms donates shoes, however, no factories appear to be planned.

32. O'Connor (2014).

33. Rupp and Banerjee (2014).

34. Merritt et al. (2010) provide a wonderful overview of the research on moral self-licensing. Of note, self-licensing occurs even when all that a person does is make a public statement of good intention, which is particularly relevant for Toms Shoes and other purchases where apparent proof of goodness is publicly visible.

35. To be clear, I'm not against capitalism. Capitalism is a terrific economic engine, and the developing world could benefit from more for-profit companies. (One problem with firms like Toms is that the owners are rich-world people, while their workers are not.) But capitalism on its own concentrates wealth (and therefore power) in the hands of a few, as so many have noted, from Karl Marx to Thomas Piketty (2014). Other forces are needed to spread growth widely, whether it's cooperatives, unions, progressive taxation, universal provision of basic needs, private charity, or a combination of these and other factors. Social-enterprise hype glorifies market mechanisms and therefore crowds out important approaches that come with few extrinsic rewards. We need more of what liberation theology calls a "preferential option for the poor" (Farmer 2005, p. 139).

36. Franzen (2010), p. 439.

37. Fisher (2012).

38. McNeil (2010).

39. UNESCO (2012). That still leaves over 50 million children out of school, though.

40. International Committee of the Red Cross (2014).

41. Richard Davidson is a leader in the field of affective neuroscience, which seeks out the physiological underpinnings of emotion. Two papers discuss the link between activity in the prefrontal cortex and subjective positive mood: Davidson (1992) and Davidson et al. (2000).

42. Richard Layard's (2005) book is a superb, easy-to-read introduction to the modern economist's view of happiness.

43. Cobb et al. (1995).

44. Sen (2000), p. 14. *Development as Freedom* makes a powerful case that socioeconomic growth comes through the provision of freedoms and capabilities. Ultimately, though, the underlying philosophy provides an apology for liberal free-market democracy, with no discussion of the responsibilities that individuals must also cultivate. This book's Part 2 is a response to Sen that could be called "Development as Wisdom."

45. Seligman (2002), Gilbert (2006), and Haidt (2006) are among psychology's foremost scholars of happiness and "positive psychology," and each brings his unique perspective. Rubin (2009), who is not a psychologist, tried a wide range of happiness tips in her own life for over a year. All of these authors acknowledge the importance of character and virtue as a cause of happiness, but in a demonstration of exactly the tendency I critique in this section, they focus more on tricks to improve present mood or to reinterpret the past. They spend surprisingly few pages on character and virtue: Seligman has an 8-page chapter on "Renewing Strength and Virtue"; Gilbert spends 3 pages explaining that virtue and happiness are different; Haidt's "Felicity of Virtue" section is 25 pages out of 244; Rubin, surprisingly, spends the most time on virtues in the form of everyday habits. Still, many of those habits are again habits to improve present mood, and as Hoffman (2010) notes, she says little about the upbringing that allowed her to have a blessed life.

46. *Wall Street Journal* (2009).

47. Bentley (2012).

48. Obama (2013).

49. See, for example, Perry (1990), pp. 183–184. The original story has a cicada instead of a grasshopper, but I use the grasshopper here because it is more familiar to American audiences.

50. Ed Diener has led much of psychology's attempts to define and measure subjective well-being. Two of his coauthored works offer great summaries of what is known: Diener et al. (1999) and Diener and Biswas-Diener (2008).

51. Lyubomirsky (2007) epitomizes the positive psychology movement, which, though it is based on good science, seems primarily concerned with mental tricks to uplift one's mood, rather than the hard work of laying the groundwork for a happy life. Admittedly, I have cherry-picked from her book to make my point, but the unpicked fruit is not that different. Her book, and positive psychology as a whole, has a tendency to neglect important virtues in favor of easy lessons drawn from the latest research studies. For example, Lyubomirsky devotes roughly the same amount of space to "expressing gratitude" and "practicing acts of kindness," though the latter seems dramatically more involved and more likely to increase happiness in the world. It also doesn't help that at every turn, she takes pains to mention how little effort happiness requires–acts of kindness can be "small and brief," and "many of the happiness activities do not actually require you to make time." But isn't this expectation of happiness without investment one of the problems of modern society's widespread unhappiness?

52. For a scathing attack on positive psychology and superficial recommendations for happiness, see Barbara Ehrenreich (2009). She chronicles her exasperation

with the Pollyannaish positive psychology she encountered during her battle with breast cancer.

53. Wikipedia (n.d.), "Don't Worry, Be Happy," http://en.wikipedia.org/wiki/Don%27t_Worry,_Be_Happy.

54. I don't mean to be unsympathetic to people who can't pay their rent despite doing everything they can to make a decent living; nor am I denying structural causes of poverty. Some social circumstances are nearly impossible to make work. My point, rather, is that there is no simple path to happiness, and simply redirecting our aim toward happiness doesn't in and of itself address the cause of unhappiness. If anything, it can be counterproductive by drawing our attention to short-term fixes rather than to long-term foundations.

55. The Internet has amplified both our penchant for catchy fake quotations and our ability to verify actual sources. Variations of this quotation are often attributed to Albert Einstein, but thanks to O'Toole (2010), I was able to trace its true source to sociologist William Bruce Cameron (1963), p. 13.

56. The United States grew to be a major economic power well before we were able to measure GDP. In the 1930s, the economist Simon Kuznets architected the first system of national income accounts. Since then, GDP has taken on a life of its own in exactly the ways that Kuznets cautioned against. A good account of his warnings and our failure to take them into account is offered by Rowe (2008).

57. Rankism–the root of all forms of discrimination and abuse of power–is nicely defined and demolished by Robert W. Fuller (2004).

58. Quoted in Fisher (1988). The quotation, though widely attributed to Genghis Khan, is probably not his (O'Toole 2012).

59. The Tech Commandments are increasingly shared by people of all political backgrounds, but they do have a decidedly libertarian flavor. George Packer (2013) notes Silicon Valley's libertarian leaning. Anyone who believes in technology and free markets alone to redeem politics and social challenges, though, needs to travel more. If you have technology, markets, and freedom without a strong state, you have Somalia.

60. Deutsch (2011) makes a scientist's case for the critical role of the Enlightenment.

61. The seeds of the Enlightenment itself might have been planted earlier–Gutenberg's press, for example, was invented in the mid-1500s, at least a century before the Age of Reason. Colonization, which contributed to Europe's economic development, began in the 1400s. Nisbet (1980) traces the idea of progress back to ancient Greek civilization. Nevertheless, it seems safe to say that elements of today's Tech Commandments were first given concrete voice during the Enlightenment–a point that is made by many whom Nisbet (1980) cites.

62. Data in this paragraph comes from the following sources. GDP: World Bank (2012a), data for 2006 GDP world and OECD member countries. Life expectancy: United Nations (2007); US Department of Commerce, US Census Bureau (1949). Democracy: Kekic (2007). Happiness: Inglehart et al. (2008).

63. US Energy Information Administration (2014a); *Economist* (2014); MarketLine (2014).

64. Estimates for this ratio are hard to track to primary sources, and few sources provide the units of comparison. The per capita figure comes from Scheer and Moss (2012). Diamond (2008) notes a ratio of thirty-two times the natural resources. United Nations Environment Programme (2011) puts the ratio at ten times by weight in natural resources. Whatever it is, developed-world consumption, and American consumption in particular, is much, much higher than developing-world consumption. Poorer people, by necessity, know how to make a little go a long way.

65. The Sarkozy quotation is from *Wall Street Journal* (2009). The reportage on Sarkozy's commission appears in Uchitelle (2008).

66. Stiglitz et al. (2009).

67. In an incident that simultaneously illustrates a cause of the technocratic crisis as well as an outcome, when I was shopping this very book to publishers, one editor told me this: "I worry that part of what makes this book so distinctive–it's evenhanded, less polemical approach–is the same thing that will make it harder for us to sell. Right or wrong, it's often those books that take a firm stance that we have the most success with." To paraphrase . . . We don't care if you're right or wrong; we'd rather have a more polarizing, less balanced perspective.

Chapter 6: Amplifying People

The Importance of Heart, Mind, and Will

1. Sawyer (1999).

2. Swaminathan (2005).

3. *Hindu* (2006).

4. Jhunjhunwala et al. (2004).

5. See, for example, Best (2004).

6. It was either Veeraraghavan et al. (2005) or (2006).

7. For more about telecenters, see Kuriyan and Toyama (2007).

8. Internet cafés typically end up catering to young men playing video games and consuming adult content. Assuming they do well as businesses, their primary benefit is to the entrepreneur in the form of increased or diversified income. But this is the same benefit that any additional line of business would have for a small entrepreneur, and success is less a function of the technology as it is the entrepreneur's skill. Those same entrepreneurs often sell an array of products and services,

but few people make a fetish of, say, using cigarette sales to increase business the way that telecenter proponents do with computers. For a comprehensive overview of telecenter research, see Sey and Fellows (2009).

9. Digital Green was modeled on another project we supported at Microsoft Research India called Digital StudyHall (n.d.).

10. Gandhi et al. (2009).

11. For an alternate description of Digital Green, see Bornstein (2014).

12. Jack and Suri (2011), Mbiti and Weil (2011), and Morawczynski and Pickens (2009) all report that the frequency of urban-rural remittances is greater with M-PESA. Mbiti and Weil (2011) and Morawczynski and Pickens (2009) also suggest that the total amount of remittances is greater. Morawczynski's (2011) PhD thesis looks at M-PESA's rise and usage patterns in depth.

13. It's very tempting at this point to suggest that Partner X become the go-between between the Internet and Partner X's constituents. Whatever pregnant mothers want to know, Partner X would look up online and relay to the mothers. But unless Partner X has health-care providers on its staff, this is naïve and dangerous. Would *you* go to a hospital where the staff members aren't trained doctors and nurses, but people who look up articles on Wikipedia and study surgery on YouTube?

14. This phenomenon is not rare. I've been to several places in both India and parts of East Africa where communities have had so many failed packaged interventions foisted upon them that they have grown cynical of outsiders coming in with yet another one. Some communities are outright hostile. Anyone committed to supporting these communities must undo the damage of earlier efforts first, before being able to meaningfully engage.

15. In a study with its partner, Voluntary Association for Rural Reconstruction & Appropriate Technology, Digital Green was found to increase annual income by 68 percent, on average, from $144 a year to $242. Some households saw their incomes double.

16. It's also possible for technology projects to build the institutional capacity required from scratch. Grameen Foundation (2014), a nonprofit I advise that seeks technological innovations for global poverty, did exactly that in its Community Knowledge Worker (CKW) project in Uganda. It identified, recruited, trained, and empowered local villagers to serve as CKWs in their communities.

17. Ramkumar (2008) includes a case study on social audits, including challenges of implementation, as written by a former MKSS member.

18. Veeraraghavan (2013).

19. "Vincent" is a pseudonym used here to protect the boy's identity.

20. Gamification is a hot trend among tech-minded social activists, but it turns out to be incredibly difficult to design games that people voluntarily play

which are also educational or productive. The essence of the problem is that it's difficult to hit two birds with one stone. It's very hard to design a compelling game, and it's very hard to design a good educational app, so it's extra hard to design a compelling educational game. Any given educational game is inevitably less compelling than the best fun-only games. Thus, the fun-only games tend to thrive and win out.

21. Organisation for Economic Co-operation and Development (OECD) (2014b), pp. 305, 382.

22. International Math Olympiad (2014).

23. OECD (2011), p. 230; OECD (2013b), p. 174. The "ninth-worst in educational disparity" statistic is based on gaps in PISA math scores between students in the top ninety-fifth percentile and the bottom fifth percentile in socioeconomic status.

24. Duncan (2012).

25. I want to be clear that I'm not arguing against, say, programming classes for underprivileged teens. The important part of such programs, though, is that they apply quality resources preferentially for those with the least advantages–*not* that they involve technology. Thus, mass distribution of tablets is pointless, because it's not quality education in itself. But afterschool arts programs for the children of poor families would be good, even if it doesn't involve high-tech.

26. Warschauer (2006).

27. The OECD's (2011) recommendations for the United States span twenty-eight pages and focus on culture, teacher capacity, administration, and spending policy (pp. 227–256). There isn't a single word about computers or other technology, except to illustrate resource differences between rich and poor schools.

28. Bilton (2014).

29. Shirky (2014).

Chapter 7: A Different Kind of Upgrade

Human Development Before Technology Development

1. Ratan et al. (2009). Positive responses like this are partly genuine, but Dell et al. (2012) also find that recipients of packaged interventions are good at second-guessing what providers want to hear.

2. There is an endless debate in the international development community about whether providing people with entertainment is worthwhile or not. See, for example, Arora and Rangaswamy (2014). Certainly, they increase momentary pleasure, and it could be argued that escape offers a palliative for an otherwise difficult life. And in any case, it seems wrong to prohibit or hinder entertainment. But it's another thing to spend scarce resources for purposes whose long-term

contributions to well-being are fleeting, and which could simply be sedating people into accepting unacceptable conditions (entertainment is the opium of the masses?). At the very least, if entertainment is the primary goal of a packaged intervention, proponents should advertise the goal as such, not fall back on it as a last-resort benefit for otherwise unimpactful projects.

3. This was especially true in India, where the pay difference between menial work and even the least demanding office work can easily be a couple of zeroes.

4. Ratan et al. (2009).

5. Drexler et al. (2010).

6. Banerjee et al. (2011).

7. Mitra and Arora (2010).

8. "Learned helplessness" is a psychological phenomenon first described by renowned psychologist Martin Seligman. Seligman and Steven Maier (1967) first conducted experiments with dogs that showed that if dogs were inflicted with prolonged electric shocks that they couldn't escape, many (though not all) stopped bothering to try to find ways to avoid the shock altogether. Notably, the learned helplessness persisted even after the dogs were offered an exit. Corresponding tendencies have been found in human beings, particularly with certain instances of depression (Seligman 1975).

Anthropologist Oscar Lewis (1961), who observed such traits in poor communities in Mexico, the United States, and elsewhere, believed they were social adaptations that were both a result and a cause of impoverished conditions. His notion of a "culture of poverty" was politically hijacked in America as a way to blame poor communities for their own plight, but Lewis meant it in a very different way. What he saw was that congenital poverty teaches lessons that are useful for survival, but not necessarily optimal for escape. So, for example, under conditions of extreme poverty, whatever effort a farmer puts in on the farm, other factors, such as pests, bad weather, or corrupt bureaucrats, might have more influence on his income. The circumstances don't encourage personal initiative. Or, where there is an urgent demand to put a meal on the table today, it's hard to learn the value of saving for the future. Or, if unjust authority figures are particularly ruthless, it might be safer to accept one's situation than to expend one's energy rocking the boat. The intention to help one's future self can be snuffed out under extreme hardship. Lewis writes, "The subculture of poverty can be viewed as attempts at local solutions for problems not met by existing institutions," and "the culture of poverty [has] a counter quality and a potential for being used in political movements aimed against the existing social order." A good review of the issues occurs in Small et al. (2010), who also conclude that careful, sensitive study of culture's role in poverty is merited.

9. Singer (2011).

10. Controversially, some differences between groups that are often explained as differences in culture or personality could be explained as different breadth of intention. For example, if superior intention correlates with concern for larger circles of life, then the radius of concern is one measure of intrinsic growth. Isn't it wiser for a society to honor women's rights as well as men's rights? To seek the benefit of people in other groups or nations as well as one's own? And even to be sensitive to animal suffering as well as to human suffering? As Jeremy Bentham (1789 [1907]) noted, "The question is not, Can they reason? nor, Can they talk? but, Can they suffer?"

11. Bourdieu (1979 [1984]). Sociologist Pierre Bourdieu's discourse on cultural capital is part description and part political critique. His core claim is that various forms of social and cultural capital enforce class barriers, and that they are propagated by education and other social structures that have historical determinants. I borrow his ideas here without the social critique–middle-class cultural capital is important for anyone wanting a middle-class life. Bourdieu often rambles, so his interpreters have been helpful. See, for example, Grenfell (2008).

A similar argument is made by sociologist Annette Lareau (2011), who follows parenting styles in different households. What she finds are stark differences between working-class families and middle-class ones, leading to what she calls a "transmission of differential advantages": Better-off families inculcate habits of the better off in their children; working-class families inculcate habits of the working class. Lareau cites Bourdieu and shares his social critique, but the problem is less that class advantages and disadvantages both propagate across generations (which has the positive benefit of ratcheting any gains that families make) as that we don't have social systems in place to help the less-privileged children rise beyond their heritage.

Earlier drafts of this book contained a chapter on intergenerational transfer of intrinsic growth. I may make a version of it available at my website (http://www.geekheresy.com).

12. Carol Dweck (2007) is a leading psychologist whose research shows the value of a "growth mindset" over a mindset that values traits that are hard to change. Her book's back cover says that it's something that "all great parents, teachers, CEOs, and athletes already know." That is, those with discernment intuit this without the research. Mueller and Dweck (1998) show how childrearing is better served by praising effort (thus leading to a growth mindset) than by praising ability.

13. Differences along these lines lead to vastly different levels of attainment even in the developed world, simply because they act as barriers between social classes, as Bourdieu (1979 [1984]) emphasizes. For example, personal initiative and effort are underappreciated with some consistency in communities that rarely

witness upward social mobility. Oscar Lewis (1961) cites "resignation and fatalism based upon the realities of their difficult life situation" as one of many characteristics he observed in impoverished communities.

Coleman et al. (1966) report on the status of US education, where they found that children from disadvantaged backgrounds were more likely to believe that luck matters more than individual effort. The report was a landmark study of the state of American public education, particularly with respect to its effectiveness on different racial groups. It confirmed the importance of parental educational status and documented the relatively small role that schools played to equalize incoming disparities in academic achievement among students. Among its most interesting findings was a disparity in how different racial groups felt about their sense of control over their lives. In response to statements such as "Good luck is more important than hard work for success," and "Even with a good education, I'll have a hard time getting the right kind of job," disadvantaged groups were more likely to agree compared with the white, middle-class majority. The report also strongly suggests that these differences are a factor of the environment–upbringing and education–and argues against segregation. The verbatim summary:

The responses of pupils to questions in the survey show that minority pupils, except for Orientals, have far less conviction than whites that they can affect their own environments and futures. When they do, however, their achievement is higher than that of whites who lack that conviction. Furthermore, while this characteristic shows little relationship to most school factors, it is related, for Negroes, to the proportion of whites in the schools. Those Negroes in schools with a higher proportion of whites have a greater sense of control. Thus such attitudes, which are largely a consequence of a person's experience in the larger society, are not independent of his experience in school.

A similar emphasis on luck has been reported in other cultures, as in the notes that straddle this one. Research such as Henrich et al. (2004) and Jakiela et al. (2012) find similar luck-focused beliefs common among poor communities, such as in the Peruvian Amazon and rural Kenya. Of course, the belief that effort goes unrewarded is often a survival mechanism for conserving energy and minimizing despair, but it is a self-defeating prophecy that reaffirms the status quo.

14. "Practical wisdom" as defined by Schwartz and Sharpe (2010) is very close to the concept of discernment that I am defining in this chapter.

15. With regard to individuals, there is a rich line of research in the psychology of self-control (explored under various names, such as executive function, self-discipline, self-regulation, delay of gratification, and willpower), as well as in its pathological absence (such as akrasia, the breakdown of will, self-defeating behavior, and, in an extreme form, addiction). Academic experts sometimes make fine distinctions between these terms, but the concepts are closely related. Among those who

champion the primacy of willpower are Walter Mischel, George Ainslie, and Roy Baumeister. Mischel is best known for his "marshmallow experiment" which demonstrated that young children who were able to delay gratification by giving up an immediate reward for a larger reward later grew up to be more successful in school and life than their peers who were not. See Shoda et al. (1990) and Mischel and Shoda (1995). Baumeister and his colleagues confirm that self-control is a predictor for better health, education, and employment, and further find that greater amounts of it as a character trait appear to confer a consistent advantage in life. See, for example, Tangney et al. (2004). Baumeister and Alquist (2009) also argue that self-control is an unmitigated good in the sense that having more capacity for it has no drawbacks (e.g., having more self-control doesn't mean overusing self-control). A very readable summary of Baumeister's findings occurs in Baumeister and Tierney (2011). Ainslie (2001) looks at the dark side of lack of willpower. He is known for proposing "hyperbolic discounting" to model how people consistently prefer a near-term reward far more than they ought to, at least as expected by standard economic models of time discounting.

Psychologist Roy Baumeister, whose seminal work on self-control has reinvigorated modern science's interest in the concept of willpower, finds that all effortful human activity–e.g., dieting, concentrated thinking, physical exertion, and emotion regulation–draws on the same physiological reservoir of will, one linked to glucose in the bloodstream. Meanwhile, studies confirm that low self-control leads to "compulsive spending and borrowing, impulsive violence, underachievement in school, procrastination at work, alcohol and drug abuse, unhealthy diet, lack of exercise, chronic anxiety, explosive anger" (Baumeister and Tierney 2011). Baumeister distinguishes between "trait self-control" and "state self-control." If self-control is a reservoir of water, trait self-control is the total capacity of the reservoir, and state self-control is how much water is left at any given point. With respect to individual self-control, we should make efforts to raise trait self-control.

16. Emotional intelligence is the concept popularized by Daniel Goleman (1995). To the degree that emotional intelligence is about intention, discernment, and self-control, it overlaps greatly with intrinsic growth. However, Goleman's concept of emotional intelligence also includes traits, such as the ability to emotionally empathize, that are not strictly necessary for intrinsic growth. It's possible (if difficult) to be wise without having the emotional sensitivity that Goleman highlights. Economist James Heckman's (2012) usage of the phrase "noncognitive traits" also overlaps considerably with the idea of judgment and self-control, but his definition lacks the element of intention.

17. Intrinsic growth is internal to, and under the partial control of, the person or the society. Heart, mind, and will are neither external advantages nor

purely inborn talents, even though circumstances, genetics, and maybe even epigenetics can play a part in forming them. How healthy you are depends on your genes and the larger environment, neither of which you can personally control. Yet, you do have within your control the ability to gain the intention to have good health, the discernment to choose nutritious foods, and the self-control to go for a daily walk.

18. Oppenheimer (2003) and Toyama (2011) both have strong things to say about television's poor performance in education. Wilbur Schramm (1964), considered the father of communication as a field of social science, illustrates some of the high expectations of TV for international development in the 1960s.

19. See, for example, Polgreen and Bajaj (2010).

20. Peterson and Seligman (2004) provide a reference book that identifies twenty-four character strengths clustered into six virtues that the authors found to be valued by cultures across the globe–knowledge, courage, humanity, justice, temperance, and transcendence. Assmann (1994), Takahashi and Overton (2005), and Yang (2001) provide additional discussion of cross-cultural issues in wisdom.

21. See Paulhus et al. (2002) and Takahashi and Overton (2005) for two such lists. Paulhus et al. (2002) additionally demonstrate that people distinguish clearly between wisdom, intelligence, and creativity.

22. I use well-known examples, but it's very possible to be capable of high intrinsic growth without being notable or famous. I used to volunteer at a hospice, and a few of the nurses who worked there seemed like pure angels–they were deeply compassionate, highly capable, and worked long shifts that were frequently visited by death without losing their heart or their cool . . . and all with little praise or recognition. They were models of heart, mind, and will.

23. I think of the three pillars as what are called "basis vectors" in vector mathematics. The bases of intention, discernment, and self-control span the total vector space of causes of virtuous activity. I should mention, incidentally, that I don't necessarily mean intention, judgment, and self-control to correspond to physiological or psychological constructs. They are philosophical concepts around which to organize a theory of social change. It may very well turn out that good judgment, for example, is a complex combination of twenty-three separate mental faculties, seven of which also underlie good intentions. So, while the three pillars are conceptually independent, in our brains the wiring may be interconnected. For example, there's a large and convincing body of research suggesting that greater self-control leads away from criminality and toward prosocial behavior, which means that expressed intentions themselves might change through greater self-control (e.g., Ainslie 2001). What we often call enlightened self-interest might be a case of selfish intention combined with discernment that leads to less selfish intention in practice. The psychology matters to the extent that we need to

understand it to nurture the traits; but for the sake of social change, it's the final, expressed traits that matter. Whether you volunteer your time to charitable causes out of empathy or out of cold, long-term self-interest matters less than the fact that you volunteer–the expressed intention is similar.

24. Of course, good health also requires other things, ranging from genes and luck to the right medical technologies, but these other factors are largely outside of individual control. And if they're a part of individual control, they will fall into something covered by intention, discernment, and self-control.

25. See Bloland et al. (2012) for how the US Centers for Disease Control and Prevention views the importance of health systems. The 2014 Ebola crisis was an object lesson in health systems, since the disease has no known cure as of this writing, yet there was a huge difference in death rates between those who were treated in Africa and those who were treated in America. Global health luminary Paul Farmer noted that there was a "know-do gap" between what we know we should be doing and what actually gets done; much of that gap is due to weak health systems (Achenbach 2014).

26. I-TECH (2011).

27. Bendavid and Bhattacharya (2009).

28. Walensky and Kuritzkes (2010).

29. Centers for Disease Control and Prevention (2008); Colindres et al. (2008).

30. For another such example in public health, see Toyama (2012), which is about Aravind Eye Hospital and its emphasis on staff and organizational intrinsic growth.

31. I fail in this book to sufficiently stress collective action and its critical role in social change. That is partly because of limits of page count and partly because of the incredible difficulty of telling collective action stories without highlighting individuals. Based on what I've witnessed, however, unified group action is the single most effective way to address abuses and imbalances of power. Chapter 10's Pradan assists self-help groups to engage with local politics for this reason. The Egyptian revolution was more than anything a story of collective action. Farmer unions and cooperatives support their members more effectively than any farming household can do on its own. But collective action again requires group intrinsic growth–there is no effective collective action unless the group has enough heart, mind, and coordinated will.

32. Tara Sreenivasa's story is based on my interviews and email exchanges with her as well as a paper questionnaire distributed to Shanti Bhavan students who were in the eleventh grade in 2009. Dmitry Kogan, an intern who worked with me to conduct a study of English-language learning across several schools, administered the questionnaire.

33. This phrasing is used in the Constitution of India (2011), Part XVI, Clause 340.

34. Detailed descriptions of Shanti Bhavan are provided in George (2004). Thomas Friedman (2004) wrote about Shanti Bhavan in *The World Is Flat*; his daughter was a volunteer.

35. George (2004) is part memoir and part social critique. It describes George's extensive efforts to build Shanti Bhavan and other institutions and programs–a medical center, a campaign against lead poisoning, a journalism school, etc.

36. Psacharopoulos and Patrinos (2004).

37. Mandela (2003).

38. Citations have been removed from this quotation from Patrinos (2008). See also Psacharopoulos and Patrinos (2004).

39. Education is close to being a silver bullet, but it's not quite: Because the human qualities essential to an effective education aren't easily replicable, good education as a whole cannot be a packaged intervention. Textbooks are packaged interventions; laptops are packaged interventions; school buildings are packaged interventions; laws for mandatory schooling are packaged interventions. But good education itself is not. For tremendously inspiring stories of the value of girls' education, see Kristof and WuDunn (2009).

40. There are scholars who question the value of education. Some cast doubt on any correlation between education and national growth. Benhabib and Spiegel (1994) and Pritchett (1996) argued that additional years of education do not correlate with economic development at a national level. But research since then, for example, by Krueger and Lindahl (2001), has cast doubt on the quality of data from which Benhabib, Spiegel, and Pritchett worked. Even as Pritchett himself offered, school enrollment doesn't necessarily translate to good education. Obviously, educational quality is what matters, not whether a child has technically enrolled in school.

A related, more realistic problem is that education by itself may have little economic value in contexts where jobs are limited. See, for example, Bhide and Mehta (2004), Deininger and Okidi (2003), Krishna (2010), and Scott (2000). But economic value isn't everything. There's still an argument to be made that education makes it more likely for people to demand changes to a system limiting their opportunities. Friedrich Engels (1844 [1968]), p. 125, wrote that "the middle class [who own the means of production] has little to hope, and much to fear, from the education of the workers." The French and American revolutions were fanned by deep thinkers writing pamphlets against monarchy. A cleverly titled book called *The Dictator's Handbook* notes that "no nondemocratic country has even one university rated among the world's top 200" (Bueno de Mesquita and Smith 2011, p. 109). The authors continue, "Highly educated people are a potential threat to autocrats, and so autocrats make sure to limit educational opportunity."

41. There are occasional debates about how much luck and other factors (e.g., skill, effort, personality) contribute to various ends (see, for example, Frank 2012). However, while these questions have academic value, precise measures are not important for the sake of practical action. In fact, for certain things, better outcomes can result if you believe (or act as if you believe) that something that is factually wrong or exaggerated. Optimists are less realistic than mild pessimists, for example, but they're more likely to take the risks necessary for success and more likely to be happy (Seligman 2006). This is true of any quality you have control over when it occurs in opposition to luck. It's interesting, for example, that a lot of self-made people emphasize how important it is to make your own destiny, and belittle the role of luck. What that proves is not that their model of the world is accurate, but that personal success is more likely to come if you *act as if* you believe–correctly or not–that it's all up to you.

42. Coleman (1966). The Coleman Report was controversial when it came out because the effort/luck difference was found to cut across racial lines, which was no surprise then because of the strong correlation between race and socioeconomic status in 1960s America. Conservatives seized on the report as proof of bad culture among blacks; liberals reacted with claims of "blaming the victim." Ever since then, American society has been unable to have an intelligent conversation about culture and personal virtues. I return to this issue in Chapters 8 and 9.

43. See, for example, Elizabeth Kolbert's (2012) descriptions of a six-year-old girl from the Matsigenka community in the Peruvian Amazon acting like a very responsible adult.

44. See, for example, the Agricultural Self-Sufficient School by Fundación Paraguaya (n.d.). Also, prior to widespread formal education, apprenticeships were a common method of handing down knowledge and wisdom (De Munck et al. 2007).

45. This is a bit of a generalization about Japanese education, which, like any educational system, has its strengths and weaknesses. For example, as a result of misguided *yutori kyoiku* ("relaxed education") policies in the 1990s, Japanese student achievement fell for a while (Brasor 2001). Attempts to return to the previous state of rigor have just recently begun (Kato 2009). On the other hand, there are continual efforts to improve the way in which teachers teach, such as more recent efforts to have students think and discover mathematical algorithms on their own (Green 2014). On the whole, however, much of basic Japanese education is about rote memorization, especially compared with what I have experienced in American schools.

46. Wai et al. (2010) hypothesize that greater "educational dose" is one key to a good education. In addition to quality of education, variety of experience may be critical. Mahoney et al. (2006) provide a thorough review of the literature on extracurricular activities for children. They find that on the whole, and contrary to worries of over-parenting, a richer set of organized activities is correlated with better child adjustment and self-esteem–except when piled on to excess.

47. Pal et al. (2009).

48. Plutarch (1992), p. 50. Thanks to O'Toole (2013).

49. Prahalad (2004), p. 16.

50. Data on government expenditure per student is from the World Bank (2012c). The figure varies depending on how it's calculated, but $250 is a conservative upper bound. Using figures available or extrapolated for 2011, per capita GDP was $1,489; primary-school pupils numbered about 150 million; secondary-school pupils, about 110 million. During the past decade, and for years in which data is available, education expenditure per student as a percentage of per capita GDP was highest in 2003, when India spent 11 percent per primary-school student and 21 percent per secondary-school student. This comes to about $226 per student. However, while primary-school enrollment is 116 percent (due to children outside of the age bracket who attend school), and secondary-school enrollment is 63 percent, there are still 45 million school-aged children not in school. Thus, government expenditure per school-aged child is closer to $192. Meanwhile, using the total government budget of $47 billion for all education, and dividing it among the school-aged population (305 million), results in only $154 per student, even though this includes the budget for tertiary education.

51. Inglehart et al. (2008).

52. Shanti Bhavan has a policy of only accepting one child per family. Although this strikes some westerners as unfair, it aligns with cultural norms in rural India, where families tend on their own to invest in a single child. In fact, many Indian people would consider it less fair if more than one child per family were given the opportunity to attend Shanti Bhavan. Why should one family receive so much, when there are so many families in need?

53. I have encountered many stories of child sexual abuse in impoverished and less-educated communities. There seem to be complex reasons for this, ranging from co-sleeping family members to weaker social norms against abuse. However, this issue is all but invisible outside of those communities, and there is surprisingly little documentation of the phenomenon, either in scholarly papers or popular journalism. Two exceptions: National Coalition for Child Protection Reform (2003) discusses the domestic US context; and Resources Aimed at the Prevention of Child Abuse and Neglect (1997) discusses South Africa. Similar findings probably hold in many other countries.

54. An effect analogous to this was established by economist Rob Jensen (2012) in an experimental study in North India. In rural villages that witnessed some women getting jobs in the relatively lucrative outsourcing industry, other women were "significantly less likely to get married or have children during this period, choosing instead to enter the labor market or obtain more schooling or postschool training."

55. Awuah (2012).

56. Actually, it would take about thirty-six years and four months at a 10 percent growth rate per year. At 9 percent, it would take more than forty-one years. The poverty rate for a single household in 2014 was $11,670 (US Department of Health and Human Services 2014).

Chapter 8: Hierarchy of Aspirations

The Evolution of Intrinsic Motivation

1. Sandberg (2014), pp. 296–297.

2. Economists will point to the behavioral scientists in their midst and insist that their field knows money isn't everything. But it's telling that even the behavioral economists' favorite piece of evidence for this claim is the game of Dictator, in which one person allocates a pot for two players–a pot of money. In the most common form of Dictator, one player (the dictator) is given some cash–say, $10–along with instructions to split the money in whatever way he chooses with the second player (see, for example, Camerer and Thaler 1995). Over many experiments in various contexts, and to the jaw-stretching astonishment of mainstream economists, it has been found that people in the role of the dictator frequently choose to give the other player some of the money (Henrich et al. 2004). This fact upends the standard "rational agent" model of economics, which assumes that people are moved purely by self-interest and would therefore keep the entire pot for themselves.

3. *Oxford English Dictionary* (2013).

4. The sample survey was conducted by market-research firm Synovate. Nathalia Rodriguez Vega helped me analyze the data. The exact wording of the survey question was, "Among those things that you have some control over, what would you most like to change about yourself or your life over the next five years?" Over the years, I've found this wording to work best even in translation. (The word "aspiration" itself can be confusing or hard to translate.)

It's possible that people might not mention aspirations they have that are less than good, but if so, the responses still show that people know what a good aspiration is (i.e., the response reveals something true even if the respondent has what psychologists call "social desirability bias"). Even when similar questions are asked of gang members, mafioso, and dictators, they typically reveal positive intentions, though their means to achieve them might be crooked. See, for example, Venkatesh (2008).

5. Deci and Ryan (1991).

6. For an overview of these results, see Sheldon and Houser-Marko (2001) and Sheldon and Elliot (1999). Sheldon and Elliot (1998) also find that people persevere more at intrinsically motivated goals. The finding that the greatest gains come from following intrinsically motivated goals occurs in Sheldon and Kasser (1998).

7. This process is also confirmed by Sheldon and Elliot (1999): "Those who are progressing well in their goals during a period of time are accumulating activity-based experiences of competence, autonomy, and relatedness during that time, more so when their goals are self-concordant."

8. Richard Auty (1993) first identified and named the resource curse. Jeffrey Sachs and Andrew Warner (1999) linked the resource curse to economic stunting.

9. The "Dutch disease" was so named by *Economist* (1977), according to Wikipedia's article on "Dutch Disease" (n.d.), *http://en.wikipedia.org/wiki/Dutch_disease.*

10. Erling Larsen (2004) discusses how Norway appears to have avoided the resource curse, though there are indications that it's not fully clear of Dutch disease. Meanwhile, Norway's admirable contributions to international aid are documented in many places. Revkin (2008) notes how it upped its aid contributions during a recession.

11. Agyare (2014).

12. I met Awuah when I taught math at Ashesi in 2002, and we've had many discussions since then about university education, Ghana, and development. The stories of Awuah and Ashesi throughout this book are based on our conversations over the years as well as stories I heard from other Ashesi affiliates, including Nina Marini, Ashesi's founding vice president. More about Awuah and Ashesi appears in the following: Easterly (2006), pp. 306–307; Dudley (2009); Lankarani (2011).

13. The quotation is at best a very loose translation of Goethe from what first seems to appear in Corelli (1905), p. 31. Lee (1998) provides a fuller explanation.

14. See Easterly (2006), pp. 306–307. Awuah has otherwise won a number of honors for his work, including an honorary doctorate from Swarthmore in 2004; the John P. McNulty Prize in 2009; and the Integral Fellow Award from the Microsoft Alumni Association.

15. This oft-cited finding goes back to Isen and Levin (1972) and was popularized by, among others, Schwarz and Strack (1999). I mention this pioneering work not because its conclusions are wrong or unimportant, but because it is a prominent example that emphasizes the short-term impact of the external

environment. No such single study is a problem in itself–the problem is that such studies are increasingly prioritized (because they're easier and cheaper to run) over studies of slower-changing, internal traits and are becoming disproportionately influential in policy.

16. For the canonical exposition of behavioral economics' "nudges," see Thaler and Sunstein (2008), who popularized the term. Their notion of "libertarian paternalism" is among the gentlest conceptions of manipulation, and most of their ideas are undoubtedly worth implementing. But is that all we're going to ask of ourselves? Can't we go beyond nudging one another?

17. Psychologists' notions of personality development differ from the "personality development" that I have encountered in some social change efforts, especially outside of the United States. Psychological personality development is concerned with how human beings mature across a number of psychological attributes throughout their lives. "Personality development" in Indian development circles is about the development of soft skills and the outward expressions of education and middle-class membership. The two definitions overlap, but the latter implies a somewhat more superficial type of development than the former. There are three-month courses in India for "personality development" (that succeed at their stated goals), but few psychologists would suggest that personality development is something that can be completed in a three-month course.

18. Freud (1962) proposed a psychosexual theory that took a child through oral, anal, phallic, latent, and genital stages, each layering the libidinous id with a self-protecting ego and an angel-on-the-shoulder superego. Erik Erikson's (1950) psychosocial theory featured a series of eight crises, whose successful resolutions led to trust, will, purpose, competence, fidelity, love, caring, and, ultimately, wisdom. Jean Piaget investigated the developmental stages for logical thinking and scientific ability (Piaget and Inhelder 1958). Lawrence Kohlberg was inspired by Piaget to investigate stages of moral development (Kohlberg et al. 1983).

19. I don't claim that all such change always moves in a positive direction. Sometimes, someone of praiseworthy achievement decides to make a terrible step backward in intrinsic growth–a prominent example is Bernard Madoff, whose hedge-fund pyramid scheme cheated his investors out of billions. Nevertheless, my main point here is that forward growth and maturation is not an unusual thing. Contrary to anyone who believes that human nature is fixed and economically focused, positive human change at an individual level is common and often not at all about money.

20. See, for example, Haggbloom et al. (2002). Their ranking shows Maslow as the fourteenth most cited psychologist in introductory psychology textbooks, the nineteenth most revered by other psychologists, the tenth most eminent in a final analysis, and the thirty-seventh most frequently cited.

21. Maslow (1954 [1987]), p. 22.

22. Maslow (1943), p. 375, repeated in Maslow (1954 [1987]), p. 17.

23. Maslow (1943), pp. 388–389, repeated in Maslow (1954 [1987]), p. 28.

24. Empirical studies of Maslow's hierarchy show mixed results: Some studies debunk specific aspects of Maslow's theory, while others uphold them. On the whole, his basic insights have not met with hard counterevidence. Among the most cited critiques of Maslow is Wahba and Bridwell (1976), which summarizes research on evidence for the hierarchy of needs in organizational behavior. But they set up a poor straw man of Maslow, misinterpreting things in exactly the way described in this section. Neher (1991), Rowan (1998), and Koltko-Rivera (2006) provide a better-reasoned set of criticisms as well as links to other critical work. Maslow (1996) himself frequently reflected on his own work. Psychologists Edward Deci and Richard Ryan (1985) make a case for three flat needs–competence, autonomy, and relatedness–without ordering or sequence, but they also leave out needs that span a larger range of human experience, such as survival and transcendence.

25. A common criticism of Maslow's theory is that it failed to account for evil. I hardly claim to have a comprehensive answer, but within his framework, one way to explain bad behavior is that when people find it difficult–whether through personal inability, external conditions, or unrealistic expectations–to achieve their needs or aspirations, they act out in ways that are criminal, unethical, and even brutal. Thus, when survival is difficult, some people become savage. When achievement and esteem are not forthcoming, some people choose to cheat. When genuine self-actualization is denied, some people turn to hedonism.

26. Hofstede (1984) lobs criticism based on the misinterpretation that self-actualization is the peak of human achievement.

27. Maslow (1965), p. 45, writes, "The difference between the need for esteem (from others) and the need for self-esteem should be made very clear. . . . [R]eal self-esteem rests . . . on a feeling of dignity, of controlling one's own life, and of being one's own boss." Rowan (1998) argues for a split into two levels of the hierarchy of needs, and I tend to agree with him here. Esteem differs from achievement or mastery, and it does seem to precede the more substantive need, which might explain why so many people seem to seek celebrity through reality TV.

28. The quotation is from Maslow (1996), p. 31. Koltko-Rivera (2006) makes a careful argument that Maslow intended self-transcendence as a separate level, especially in his later years. The beginning of the split between self-actualization and self-transcendence is evident as early as Maslow (1961). However, Maslow continued to wrestle with whether self-transcendence is a separate category in itself, as is evident in his unpublished papers as collected by Hoffman (1996).

Maslow wrote frequently of peak experiences and transcendence, but he rarely used the term "self-transcendence." Maslow (1968), p. vi, contains several occurrences, but as an aspect of self-actualization.

29. One example of this critique is from sociologist Tony Watson (2008), p. 35, who writes that he fulfills his social and esteem needs before eating, or that his cousin in the army fulfills prestige needs before security. However, these cases are actually cases of advanced Maslovian development (which Maslow called "intrinsic learning" or "character learning"), where a person operating primarily out of a higher-level aspiration is more than willing to sustain a deficiency in a lower-level need, just as many people skip lunch for a job that compels them.

30. Maslow (1943), p. 375, repeated in Maslow (1954 [1987]), p. 18.

31. Maslow himself used the word "aspiration" only twice each in Maslow (1954 [1987]) and Maslow (1971).

32. Maslow (1954 [1987]), p. 35.

33. Robert Wright's (2000) conception of a positive-sum return in his book *Nonzero* is an example. Psychologist David C. McClelland (1961) made a similar case over half a century ago.

34. In *Generation Me*, psychologist Jean Twenge (2006) relates trends among those born in the 1970s, 1980s, and 1990s. She finds a generation narcissistically focused on itself, but also tending toward less prejudice and greater self-confidence. These apparently conflicting traits are consistent with a focus on self-actualization, which is at once the height of selfishness and a plateau of good judgment about how to achieve selfish ends. On the other hand, it's possible to stagnate in self-actualization, which is arguably the danger facing privileged society. This is why it's important to acknowledge self-transcendence as a further level of human growth.

35. When exactly a person feels ready to put aside each level of Maslow's hierarchy is a big question, and current psychology has no easy answer to it. The accumulation of wealth, for example, can serve to satisfy survival, security, esteem, and self-actualization needs to various extents, so success at it often engenders moves toward self-transcendence. Yet it's clear that there's no absolute level of wealth that causes the transition. Some billionaires seem stuck on nothing but expanding their empires, while some people of very modest means seem to care for little other than self-transcendent activity. A lot seems to depend on upbringing. Like many other aspects of personality, inclinations harden as we grow into adults.

36. Duthiers and Ellis (2013).

37. Bales (2002).

38. The pressing concern for survival becomes clear if you spend time with poor communities. Vivid accounts of this survival mentality occur in Boo (2012), Collins et al. (2009), and Narayan et al. (2000).

39. Butterfield et al. (1975), p. 260. John and Abigail Adams kept up a routine correspondence when they were apart. In this letter, dated May 12, 1780, John Adams writes of the well-tended gardens he visited in Paris and Versailles with admiration and a tinge of envy. He marvels at the art and architecture of France, but suggests that the duty he has to studying the science of government precludes him from going into more detail. One wonders, given the prominent role he played, whether it was strictly duty that held him or a deep, personal aspiration.

40. I don't mean to suggest that changing careers is a requirement of forward movement. I highlight Agyare and Awuah as exemplars because their intrinsic growth was marked by clearly visible milestones, but the milestones are simply signposts, not a cause or a necessary result of change. It's possible to undergo dramatic aspirational changes while holding the same job. One reason why it is so difficult to make the case for intrinsic growth is because so many of its effects are not visible, and we increasingly neglect truths not accompanied by tangible metrics.

41. I make this claim only of the dominant paradigms of economics. Economics is a broad field, so there are, of course, economists who study changes in preferences (such as Matthew Rabin at UC Berkeley), but they are in the minority, and as far as I know, no economist studies a systematic way in which people's preferences change as a result of psychological maturation. In a quest for the precision of the physical sciences, economists seek equations of human behavior based on measurable variables. Though nothing limits the complexity of mathematical models in theory, in practice the reality of scarce data, intractable mathematics, and an undeniable physics-envy favors oversimplification.

42. Sandel (2012), p. 85, compiles a series of prominent contemporary economists putting incentives at the center of economic thinking: "Economics is, at root, the study of incentives" (Levitt and Dubner 2006, p. 16); "People respond to incentives" (Mankiw 2004, p. 4). In development, Mankiw's point is reiterated verbatim by William Easterly (2001).

43. One exception is Nobel Prize winner James Heckman, who has cobbled together neuroscience and psychology to arrive at an economic model of investing in early childhood education. See, for example, Heckman (2012).

44. Anthropologists take pains to distance themselves from straw-man interpretations of cultural relativism set up by critics (e.g., Geertz 1984), and I apologize for the overgeneralization in this paragraph. A more nuanced treatment of anthropologists' relationship to development appears in Lewis (2005). Nevertheless, my experience has been that many qualitative researchers loathe the very idea of societal progress and the idea that one culture can be considered superior to another, especially in any moral sense. I sympathize with the aversion to ethnocentrism and cultural imperialism, but if progress is taboo, it's impossible to debate the best routes to

progress. It seems clear that a culture that engages in child trafficking, for example, is morally and culturally improved by ceasing it. The hard questions are not whether there can be progress or not, but what aspects of culture admit a notion of moral progress (as opposed to non-moral differences of taste or tradition), and how cultures can engage with one another on moral progress without one culture imperially imposing its own ideas.

45. Though economists and anthropologists both vehemently insist that they believe in individual agency–or free will–their agents supposedly respond rationally or intelligently to external circumstances, which again pushes the cause of different outcomes to different external conditions, not to different internal states.

Other social sciences have similar debates. Psychology has its person-situation debate, which pits internal personality against external situation as determinants of behavior. Sociologists talk about social structures versus individual agency. And in the public sphere, it's become fashionable to note the "fundamental attribution error," which says that behavior is more often a result of circumstances than of some underlying stable personality. It's obvious, though, that behavior is caused by a complex interaction of *both* internal states and external situations. Which matters more is difficult to answer in a general way. You can contrive contexts in which one matters more than the other. It's like asking whether an athlete's skill or the quality of his equipment matter more in her performance, but that depends on the sport, on the athlete, and on the range of skill and quality being considered.

Yet another attack on the validity of nurture comes from those who say that immutable conditions like genetics have more influence. Judith Harris (2009) made this argument for parenting versus other environmental factors. More recently, Bryan Caplan (2012) suggested that parenting mattered little. The problem with these conclusions is that they are looking at a relatively narrow band of human context. Go to a poor rural village in Bihar, and it's trivially obvious that nurture, education, and culture have a tremendous impact on what kind of adults people become.

In any case, what I'm arguing for is that in the context of large-scale social change, it's worth working on the internal determinants of behavior, even if they are a small contribution to any given outcome. To the extent that they are slow-changing and self-propagating, their impact will eventually be cumulative and large.

46. The quotations are taken from the *New York Times* (2012) transcript of the debate.

Chapter 9: "Gross National Wisdom"

Societal Development and Mass Intrinsic Growth

1. One negative consequence of labeling is "stereotype threat," in which people perform worse at tasks when they are anxious about confirming a

negative stereotype. Steele and Aronson (1995) were the first to demonstrate this phenomenon. Nothing I say is intended to recommend stereotyping–my whole point is that both individuals and societies can change, and the hierarchy of aspirations specifically outlines how. Stereotypes, to the extent that they color some people as being a particular fixed way, implicitly deny the possibility of change.

2. See Shoda et al. (1990) and Mischel and Shoda (1995).

3. If an adult's self-control is strongly dependent on his self-control as a child, and if child self-control cannot be entirely blamed on the child, then at least some of an adult's self-control is due to causes beyond his or her agency. In short, at least some portion of a person's intrinsic development, even as an adult, is due to forces beyond his or her control.

Philosophers such as Galen Strawson (2010) use a stronger version of the argument put forth here to conclude that no aspect of a person is ultimately decided by that individual, and therefore that free will is an illusion. Everything a person does is ultimately determined by outside forces. I believe this explanation makes perfect sense in theory, and as a description of reality, it seems correct. (Philosophically, the real problem with free will is that the notion of a "self" that acts, and which most of us take for granted, is flawed, as Buddhism and some Western philosophers, such as David Hume (1740 [2011]) and Derek Parfit (1984), have argued. Otherwise, we must either posit a decision-making entity that is independent of physical forces–which is the same as assuming supernatural entities such as "souls"–or intentionally circumscribe some portion of our minds as being an internal force that we artificially consider separate from the rest of the universe.)

In any case, blame and attribution remain useful as social forces. Even in a world that is fully determined, blame can serve as an external social force that encourages people to act positively. This conclusion differs from some of the confused thinking about blame as put forth by philosophers such as Barbara Fried (2013) and Strawson, who suggest that because all behavior is externally influenced, no one should ever be blamed for anything. The commonsense notion that adults are more blameworthy than children is socially valuable; yet we also have to keep in mind that they are not *entirely* blameworthy. (Mischel himself argues that self-control is both predictive and malleable.) That's the basis from which I proceed in this section and elsewhere.

4. See, for example, Kraus et al. (2010), Piff et al. (2010), and Piff et al. (2012).

5. The question of whether external conditions or individual actions matter more is unanswerable outside of specific contexts. It's easy to imagine two

individuals with such different temperaments that they would act differently under just about any circumstance. That example would suggest that individual actions matter more than external conditions. Yet it's also possible to imagine two individuals who were raised under such different external conditions that one becomes a serial killer and another a saintly hero. That example suggests that external conditions matter more than individual actions. The fact is that both matter, and trying to help either change in a positive direction is worthwhile.

6. In autocratic societies, some decisions are made by individuals, but even then, a leader rules with some complicity of the people. Even under the most extreme conditions, people–especially collectively–have a choice, even if it is an improbably difficult choice. There's always the stark option, as they say in New Hampshire, to "Live free or die."

7. Actually, although he never visited India or China himself, Weber devoted a book each to their religions. He found both inexpedient as cultural foundations for economic growth, because neither provided a moral backing for entrepreneurial effort. Confucianism stressed individuals' harmony with their social context (Weber 1915 [1951]), and Indian religions stressed the "unalterability of the world order" (Weber 1916 [1958], p. 326). Weber could be forgiven for coming to these conclusions because this was in the early 1900s when neither China nor India were the economic powerhouses they are today. In any case, his errors suggest that what matters is not a fixed aspect of culture, but something mutable.

8. Weber (1904 [1976]).

9. *Hindu* (2014). In 2008 there were similar levels of excitement for the *Chandrayaan*, which made India only the fifth country in the world to reach the moon (Indian Space Research Organization 2008).

10. Wallis (2006) distinguishes between "systematic corruption"–where politicians bend government and tyranny is the dominant fear–from "venal corruption"–where money bends government. Wallis's research finds that systematic corruption was all but eliminated in the United States by the 1890s and that today's corruption worries are almost entirely of the venal type. Many developing-world countries today, however, continue to contend with systematic corruption.

11. Common patterns of progress do not mean that cultures lose their tremendous diversity–today's Japan is a very different place from today's United States, and again different from today's India. But many things about the way modern-day Japan differs from the Japan of a hundred years ago are similar to the ways that the United States differs from the United States circa 1900, or to how

rich, urban India differs from poor, rural Hindustan. A rich set of qualitative characteristics tend to change together, if not in lockstep.

In fact, archaeologists and anthropologists often use other stage typologies to discuss societies across a larger timescale: For example, Elman Service's (1962 [1968]) well-used typology classifies human societies into bands, tribes, chiefdoms, and nation-states. Critics of such theories have paroxysms at any mention that there could be commonalities in the histories of divergent civilizations. But if overgeneralization is a bad habit of some disciplines, undergeneralization is as bad a habit of others. Unless one is wearing the blinders of certain academic ideologies, it's hard to miss historical commonalities across nations as they develop.

12. Coontz (2014).

13. Hegewisch and Williams (2013).

14. Center for American Women and Politics (2014).

15. Economic contrarian Ha Joon Chang (2010), p. 35, writes, "The emergence of household appliances, as well as electricity, piped water and piped gas, has totally transformed the way women, and consequently men, live. They have made it possible for far more women to join the labour market." This interpretation assumes that household chores should be a women's job. The real question isn't whether some technology liberated women from household work, but how changing social norms have changed the ways in which we think about gender and work. (To be fair to Chang, he doesn't deny the importance of other social causes. He also cuts the impact of the Internet down to size in comparison to home appliances.)

16. United Nations Development Programme (2013). Ninety-six percent represents 80 out of 83 countries for which there was data in both years. A similar report by the World Economic Forum (2013) tracked global gender disparities with respect to economic participation, educational attainment, health and survival, and political empowerment. Between 2006 and 2013, it found increasing overall gender equality in 95 out of 110 countries (86 percent).

17. Inglehart has overseen the collection of what is undoubtedly the world's most comprehensive and rigorously collected dataset of subjective values. The analysis he and his colleagues did drives toward a unifying theory that bridges individual psychology and societal change. As influential as Inglehart is in some circles, his work deserves to be much more widely known. Space does not allow me to do his work justice. A superb introduction to his work is Inglehart and Welzel (2005).

18. The World Values Survey maintains an active website with all of its surveys, data, and a lot of academic analysis (www.worldvaluessurvey.org/). For example, questions from the 2005–2006 wave of the World Values Survey (2005)

ask on a ten-point scale, from "not at all" to "a great deal," how much freedom of choice or control subjects feel they have over their lives. Another series of questions asks respondents to rate family, friends, leisure time, politics, work, and religion on a scale of "not important at all" to "very important."

19. The bulleted points are excerpted from World Values Survey (n.d.) and slightly edited for readability. Also on the site is the Inglehart-Welzel cultural map, which shows how culturally similar countries cluster on their two dimensions. The map is explained fully in Inglehart and Welzel (2005).

20. There is no consensus explanation for what causes national development, but it's widely accepted that civilization itself is predicated on agriculture, since it requires a military class, which leads to ruling and leisure classes. And in more recent history, there is considerable evidence that some threshold of agricultural productivity is a strong predictor of future economic development. See, for example, Sachs (2005), pp. 69–70.

21. It's easy to romanticize agrarian life, lived as it is in the beautiful countryside and at the pace of the seasons. But the poor smallholder farmers I have met–as opposed to wealthy people who can afford other lifestyles, but think owning a farm is fun–all voice a desire to have less taxing, more predictable type of work. There are over 50,000 rural-to-urban migrants in India per day (Indian Institute of Human Settlements 2012). The world crossed a milestone in 2009, when, for the first time in history, more people lived in cities than not (United Nations 2010).

22. Inglehart and Norris (2003), p. 159.

23. Inglehart and Welzel (2005), p. 139.

24. Ibid., p. 33.

25. Maslow (1954 [1987]), p. 17.

26. Inglehart and Welzel (2005), p. 37.

27. Maslow's hierarchy maps roughly not only to Inglehart's values scheme, but also to various progressions in other social sciences. There is probably a single psycho-sociological theory that underlies them all, much the way that the biology of blooming flowers can explain the changing colors of a meadow in spring. A correspondence map of various staged theories of human development is provided below. The bottom four are individual; the top four are societal. The theories don't have a one-to-one correspondence, so the matching is approximate and indicated by similar horizontal position. Brackets indicate rephrasings of the original term. Italics indicate stages I added to complete each spectrum. References are Rostow (1960); Bell (1999); Florida (2002); Inglehart and Welzel (2005); Fowler (1981); Kohlberg et al. (1983); Fiske (1993)–"Rational Legal" is Pinker's (2011) rephrasing of what Fiske called "Market Pricing"; and Maslow (1943, 1954 [1987]).

Types of Governance	*Primitive / Chaotic*	*Autocratic*	*Nationalistic*	*Bureaucratic*	*Responsive*	*Sophocratic*
Rostow's Stages of Growth	Traditional	Preconditions for Take-Off	Take-Off	Maturity	[Self-Fulfillment]	*Philanthropic*
Bell and Florida's Class Society	Agricultural Class		Industrial Class	Service Class	Creative Class	*Compassionate Class*
Inglehart's Modernization Values	Traditional / Survival		Secular Rational / Survival		Self-Expressive	*Altruistic*
Fowler's Stages of Faith	Mythic-Literal		Synthetic-Conventional	Individuative		Conjunctive-Universalizing
Kohlberg's Moral Development	*Instinctive*	Heteronomous	Individualistic-Instrumental	Impersonal-Normative	Social System	Rights-Welfare Universalizing
Fiske's Relational Modeling	Communal Sharing		Authority Ranking	Equality Matching	[Rational Legal]	*Other Caring*
Maslow's Hierarchy	Survival	Safety / Security	Esteem	Achievement	Self-Actualization	Self-Transcendence

28. Bhatt (2003).

29. Saxenian (2006).

30. These quotations are taken from Florida (2002), pp. 77–80.

31. Florida (2002), p. xiii.

32. Ibid., pp. xiv, 74.

33. Ibid., p. 101.

34. Ibid., p. 88, quoting open-source advocate Eric Raymond in Taylor (1999).

35. Ibid., p. 92.

36. Inglehart and Welzel (2005), p. 159. See also Inglehart and Welzel (2010).

37. This debate was articulated as early as the 1960s (see, e.g., Weiner 1966). In contrast to Inglehart, Guido Tabellini (2008) argues that more open institutions lead to more open values. For example, the attitudes of third-generation Americans are linked to the degree of autocracy present in their grandparents' home countries. It seems clear that the causality can go in both directions, with factors mutually reinforcing one another. One thing both Inglehart and Tabellini agree about is that individual attitudes matter. Tabellini (2008) writes, "To explain some political outcomes or the functioning of bureaucratic organizations, we may have to go beyond pure economic incentives and also think about other factors motivating individual behavior."

38. Jones (2006); National Center for Charitable Statistics (2010).

39. Blackwood et al. (2012); Corporation for National and Community Service (2006).

40. That's an inflation-adjusted 157 percent increase over the $130 billion in 1973 giving (Giving USA 2014). A nice overview by Perry (2013) reports, however, that giving as a percentage of GDP has remained stubbornly at around 2 percent since the 1970s. It peaked in 2001 at 2.3 percent.

41. Higher Education Research Institute (2008); Eagan et al. (2013).

42. Luntz (2009).

43. The Higher Education Research Institute (2008) survey also shows that the median income of families of college students is 60 percent greater than that of all US families. This fact, together with the group's inclinations toward "helping others," suggests a correlation between family income and degree of Maslovian growth, as hypothesized. Also, as Maslow might have predicted, the global recession since 2007 appears to have caused an ebbing of interest in social causes, because external conditions pulled many back to security needs.

44. Bornstein (2004), p. 4.

45. Salamon et al. (2013).

46. Organisation for Economic Co-operation and Development (2010a).

47. Development Initiatives (2010).

48. Higher Education Research Institute (2008) shows that while "helping others" is relatively higher on the agenda now compared with the immediate past, it still trails "raising a family" and "being well-off financially," which were stressed by 75.5 percent and 73.4 percent of the sampled students, respectively.

49. I know, because it takes one to know one.

50. Pinker's (2011) book is pure genius. It brings together a compelling array of information about historical (and counterintuitive) declines in violence, prejudice, and bad manners and provides carefully reasoned explanations for them. Pinker's four "better angels" align closely with the three elements of intrinsic growth–empathy and moral reason point toward good intention, reason is discernment, and self-control, obviously, is self-control. His five historical forces, liberally interpreted, are important elements of societal intrinsic growth.

Chapter 10: Nurturing Change

Mentorship as a Social-Cause Paradigm

1. The right balance of coercion and autonomy in child-rearing is a very complex issue. Similar issues repeatedly arise in mentorship, which is the subject of this chapter. Unfortunately, I have no easy prescriptions for arriving at the right balance, except to say that good discernment is needed.

2. Tricia Tunstall's (2012) book *Changing Lives* is the basis for much of this section about El Sistema. Even if you know nothing about music, the book provides both inspiration and insight for how to create large-scale social change.

3. El Sistema (n.d.).

4. This is true however you wish to define social change, and whatever you believe the goals of social change are.

5. Pradan (2014). Pradan's impact is difficult to capture concisely because, as will become clear in this chapter, it is so varied. But its success stories are numerous and widespread. Some of them are noted at http://30.pradan.net/.

6. Pradan doesn't use the word "mentorship" to describe what it does, but its leaders agree with this chapter's description of their work and their values.

7. Many social activists like to use words such as "partnership" and "cooperation" as a way to pretend that their bilateral engagements with others occur between equals. Presumably, the intention is to avoid any neocolonialist arrogance or claims of superiority, which could lead to abuse and exploitation. But abuse and exploitation are a real danger whenever there is a real disparity in power, *whether it is acknowledged or not*. The problem isn't power, hierarchy, or status differentials; the problem is abuse. On the whole, it's better to acknowledge such disparities and to be vigilant against abuse than to pretend that disparities don't exist at all. The latter is dishonest, and it causes people to drop their guard against exploitation–as so often happens when policymakers conceive of business and trade as being inherently fair (because both sides "voluntarily" enter into the exchange), even though there is plenty of opportunity for exploitation.

Nevertheless, discernment is needed on a case-by-case basis: There are instances when insisting on making the disparity explicit can backfire, as in the case of stereotype threat, a la Steele and Aronson (1995). A strong case for accepting power, but not accepting the abuse of power, is made by Fuller (2004).

8. Pradan (2014).

9. Among qualitative researchers, hope has long been identified as a key aspect of a community's mental health, and anthropologists such as Oscar Lewis (1961) have frequently noted its absence as a trait of many poor communities. More recently, the economist Esther Duflo (2012) noted in her Tanner Lectures, delivered at Harvard University, that some groups of poor people will begin to invest in sustained effort for themselves if they are given a little hope in the form of, say, domesticated animals and carefully tailored support. It's encouraging to see economists begin to take fuzzy ideas like "hope" seriously, and I hope the scholars who first identified these issues decades ago get the credit they're due.

10. See, for example, Bell (2006) and Smyth et al. (2010).

11. As a side benefit, because mentors don't set the agenda, mentors are relieved of much of the moral responsibility should bad outcomes arise. (Not that they shouldn't try to advise against things that might bring them about.) There is a world of ethical difference between telling someone what to do and providing sincere advice when asked for it, particularly if things don't go well.

12. We should also be wary of any payment that goes to mentors. Paid mentorship can be easily corrupted into high-cost consultancies of the kind that Washington, DC's, "beltway bandits" are famous for. The higher the price, the more the "mentor" has a direct stake in something other than the intrinsic growth of the mentee. The more the "mentor" has a stake, the bigger the window for corruption. (Corruption of mission, incidentally, may be perfectly legal; but it's still corruption.) The

most obvious form of corruption among high-priced "mentors" is to place job security over mentee growth. Exactly this dynamic was exposed in 2012 when US Agency for International Development chief administrator Rajiv Shah changed the organization's policy to purchase more goods and services from local markets, and less from American ones. Groups representing beltway bandits promptly engaged a lobbying firm to fight the policy (Easterly and Freschi 2012). The best mentors offer their services free, at cost, or something very close to cost–certainly not at the maximum price that the market will bear.

13. Apostle (1984), p. 21.

14. An exception is former World Bank adviser David Ellerman (2005), who wrote a book called *Helping People Help Themselves*. His recommendations presage mine, although he doesn't use the word "mentorship." David Ellerman sent me an email while I was writing this book, and I visited him at his home in Riverside, California, soon thereafter. His book comes out of his unique experiences in development and at the World Bank, where he saw how a "church" culture of top-down edicts and unified messaging conflicted with allowing countries to determine their own course.

15. These practices are often good policies, but, strictly speaking, they aren't mentorship. However, as I note later, it's possible that within a broad framework of mentorship, incentives could be used as a way to help people come to a realization that something is of value to them. The primary difference between mentorship and manipulation as an approach is that in mentorship, the ultimate goal is a positive change in the person, not just in the behavior.

16. In the business literature, there is a rich body of work that agrees on what it takes to mentor well, at least for one-to-one mentoring relationships. I have organized this chapter based on my own experience and observations, but the recommendations do not deviate significantly from practical guides such as Brounstein (2000), Johnson and Ridley (2008), Zachary (2012), or from the academic literature (see, for example, the *International Journal of Evidence-Based Coaching and Mentoring*, with a website at http://ijebcm.brookes.ac.uk/).

Another field in which mentorship is taken very seriously is youth development. See Rhodes (2008) for a good review of the impact of youth mentorship, and *The Chronicle of Evidence-Based Mentoring* (http://chronicle.umbmentoring.org/) for related articles. A survey of over 30,000 college graduates by Gallup and Purdue University (2014) found that good mentorship in college matters much more for engagement in the workplace and overall well-being than what college a person attends.

Pawson (2004) includes links to mentorship in the context of international development.

17. The International Coach Federation (n.d.), for example, defines coaches as those who help "individuals or groups [set and reach] their own objectives."

18. For more on Farmer Field Schools (FFSs), see, for example, Davis et al. (2010), which reviews the literature on FFSs and reports on a study of FFS impact in East Africa.

19. Debates rage on about the overall impact of the Green Revolution. Critics suggest that it depleted aquifers, ruined the soil through mono-cropping, aggravated rural inequalities (amplification, yet again), and made developing-world agriculture dependent on developed-world corporations–all valid points. Here, though, I invoke the Green Revolution to discuss its efficiency with respect to its stated aim to improve yields, at which it succeeded. The Green Revolution is often described as a triumph of new technology–seeds, fertilizers, and pesticides–but it was actually a coordinated push on multiple fronts, including large investments in developing research capacity and broad-based agriculture extension. See, for example, Hazell (2009).

20. Hu et al. (2012). The Chinese extension force was 1 million members strong at its peak, but that number has since decreased. The quality of extension itself has had its ups and downs, but the government's commitment to agriculture remains strong. The most common criticism of the Chinese system is that it's top-down. Farmers are often commanded to change what they grow for the sake of national priorities.

21. Deo's incredible life story is chronicled in Kidder (2009).

22. I-TECH (2013) hosts a video where Sethu tells his story, which hints at many elements of good mentorship and its effects. Sethu is also an example of how good mentorship propagates itself.

23. Tunstall (2012), pp. 71, xi. Abreu's faith in El Sistema as social development is backed by some evidence of improved school attendance, decreased juvenile delinquency, a high rate of Sistema participants from poor backgrounds, and lower high-school dropout rates compared against non-Sistema Venezuelan teens (6.9 percent vs. 26 percent). All told, the Inter-American Development Bank estimated that every dollar invested in El Sistema was worth $1.68 in social returns, according to Lubow (2007).

24. The quotation here is taken from Pradan's (n.d.) online mission statement. Pradan is one of the wisest development organizations that I have come across. (Which is not to say that there aren't others of similar caliber that I'm much less familiar with. In India, that list probably includes SEWA, MYRADA, Seva Mandir, Gram Vikas, MKSS, Timbaktu Collective, and so forth.) Pradan's co-founder, Deep Joshi, speaks of his work in terms of "hearts and minds."

25. Packaged interventions amplify human forces to produce outcomes. If you interpret this concept in a literal, arithmetic way, your mind's eye should see a

rectangle whose width is determined by the amount of packaged interventions and whose height is determined by the strength of human forces. (In macroeconomics, the Cobb-Douglas function models economic output as the product of technology and human capital [and financial capital], so this literal interpretation is not as crazy as it might seem.) Given a particular rectangle, the optimal way to grow its area is to lengthen the shorter of the two sides. For example, if you start with a 1x5 rectangle, and have one unit to add to either side, it's better to go for (1+1) x 5 = 10 than 1 x (5+1) = 6.) Therefore, in a world obsessed with technology, it's the shorter side, the human side, that needs the boost.

26. The growth of the computing industry could be due to a similar effect within the technology world itself. It's much, much easier to develop, test, distribute, and be paid for software apps than it is to develop, test, distribute, and be paid for an efficient mechanical engine. As a result, information technology grows by exponential leaps, but automobiles (just to take an example) haven't changed much. Young people inclined toward engineering might be attracted to computing because it's the easiest route to fame and fortune. As with everything else, there's a natural tendency for civilization to warp toward the easiest tasks.

Conclusion

1. Singer (2009), pp. 3–5.
2. See, for example, American Red Cross (n.d.).
3. Easterly (2014), p. 7.
4. The thrust of *Geek Heresy* is a form of *consequentialist virtue ethics* applied to social causes. Philosopher Julia Driver (2001), who lucidly espouses this view, writes that virtue is "a character trait that systematically produces a preponderance of good." The point of consequentialist virtue ethics is to foster certain tendencies in people for the sole reason that they are likely to result in a better world. Unlike the other major categories of ethics–such as the deontological rule-based ethics championed by Kant, or the utilitarian ethics put forth by Jeremy Bentham and John Stuart Mill–virtue ethics acknowledges that knowing what's right is different from doing what's right. It focuses more on the attempt to become more capable of ethical action.

Much of Western virtue ethics, both ancient and modern, is muddled by the teleological reasoning generally attributed to Aristotle–that virtues are worthwhile because they are the best expression of what makes us human. Those who see virtue ethics this way see virtues as ends in themselves, and they often speak of human "flourishing" as having axiomatic value. Philosopher Martha Nussbaum (2011), who, together with Amartya Sen established the "Capability Approach" to international development, tends toward arguments along these lines. But flourishing in the absence of a larger goal leads perilously close to an empty high culture or a misguidedly narcissistic conception of self-actualization. More practically, neither Nussbaum nor Sen emphasize the importance of fostering the right intentions in people. They're focused on providing freedoms, but not on nurturing responsibilities.

My reading of Aristotle, incidentally, suggests that he was actually more of a consequentialist than he is given credit for. Though he does seem to have viewed virtues as noble in and of themselves, he also understood virtues to be at least a partial cause of eudaimonic happiness: "activities in accordance with virtue which play the dominant role in happiness" (Apostle 1984, p. 15). And he began his exposition in his *Ethics* with happiness as the ultimate good, probably because he meant the rest of his discussion of virtue to be in service of it.

5. The importance of health systems strengthening is concisely captured in Bloland et al. (2012). Paiva's (1977) definition of social development was the "capacity of people to work continuously for their own and society's welfare." Morozov's (2013) critique of Silicon Valley's save-the-world-with-apps mentality is excellent, but the solutions he offers are stronger in Morozov (2011). Ravitch (2011) issues a scathing critique of the "educational reform" movement in American education; David L. Kirp (2013) stresses the importance of basic management and good support of teachers. Easterly (2006) is the most constructive of his otherwise critical oeuvre; Easterly's (2014) critique of technocratic solutions is similar to mine, though our policy recommendations differ. An overview of the institutional turn is provided by sociologist Peter Evans (2005). There are many philosophers of virtue ethics, but I agree most with Julia Driver (2001). For an overview of communitarianism, see Etzioni (1993).

6. Wills (2002), p. 36. The quotation is from Madison's speech at the Virginia Ratifying Convention, June 20, 1788.

7. Lincoln (1858).

8. This needs to be said because contemporary secular society has a deep ambivalence about the idea of self-improvement, particularly in America. On the one hand, we admire self-made people, like the fictional Horatio Algers or the real-life Steve Jobs. On the other hand, these people are admired primarily for their technical abilities or their defiance of convention, not for their compassion and virtue. If anything, we chortle at sincere earnestness the way cool kids sneer at the honor society nerds. There's something about self-conscious moral striving that we belittle. The cause might be cynicism about human nature (human nature can't change); an overcompensating disillusionment with religion (which virtue reeks of); an aversion to being seen as a Goody Two-Shoes; and inability to acknowledge one's own moral failures. (As comedian George Carlin [1984] once quipped, "Anybody driving slower than you is an idiot, and anyone going faster than you is a maniac.") Even among our most empathetic literary minds, one of the worst condemnations of art

and literature is to call a work "didactic"–the faintest whiff of moralizing is considered unsophisticated and undesirable. Coolness is valued more than goodness.

It didn't always used to be this way, and there are secular cultures in which intrinsic growth is a conscious project. Benjamin Franklin (1986) listed thirteen virtues in his autobiography and went on to describe a weekly report card of virtues he made to keep tabs on his progress. Nothing could be nerdier. Mohandas Gandhi, also in his autobiography, details his "experiments with truth," in which he works out both theory and practice relating to vegetarianism, celibacy, nonviolence, and simple living. In Japan, you can see conscious attempts at virtue in small details. For example, it's common to see traffic signs with sincere admonitions to show courtesy to other drivers or to keep the roads clean. (The equivalent signage in the United States has to appeal instead to humor or threats: "Please drive safely–our squirrels don't know one nut from another." "Litter and it will hurt: $316 fine.")

9. For a catalog of these missteps in international development, I strongly recommend Easterly (2001).

10. Consider, for example, that Rachel Carson's *Silent Spring*–arguably the first milestone of American environmental awareness–was published in 1962, when US per capita GDP was $3,100 in 2012 dollars (World Bank 2012a). Indian per capita GDP today is $1,500 (World Bank 2012b). It will take at least another decade for India to reach America's 1962 levels even at 7 percent year-on-year growth.

11. Park (2014) notes this statistic, which is based on data from the World Health Organization (2014).

12. Kralev (2009).

13. If the 2014 US-China climate deal was a bit of an exception, it was exactly because President Obama brought with him American willingness to cut carbon emissions. Congressman Henry Waxman said of it, "History may look back and say this was the turning point on climate" (Parsons et al. 2014). Let's hope it sticks.

14. Figures are as posted by the US Energy Information Administration (2010) and include only CO_2 emissions from the burning of fossil fuels in 2010.

15. Francis Fukuyama (1992) contended that liberal democracy represents the "end of history"–the summit and end point of human civilization, which other nations would eventually tend toward. The thesis has been heavily criticized, not least by Fukuyama himself.

16. Asimov (1942 [1991]), p. 126. Asimov's thinking about the laws of robotics was philosophically much deeper than presented here, though none of it changes what I'm trying to say in these paragraphs. In a novel called *Robots and Empire*, Asimov (1985 [1994]) has the robot Daneel formulate a law that

supersedes the First Law: The Zeroth Law of Robotics, which says, "A robot may not harm humanity, or, by inaction, allow humanity to come to harm," allows a robot to harm a person it judges will cause far greater harm to humanity. But Daneel (and presumably Asimov) were not comfortable with this ethical can of worms. In *Foundation and Earth*, Asimov (1986 [1994]) has Daneel effectively retract the Zeroth Law by seeking out a particularly discerning human being to make a tough choice about the direction of galactic civilization. In doing so, Daneel serves as a role model for social activists–we should all be as thoughtful in allowing "beneficiaries" to determine their own fates (though one wonders whether it was sufficient for Daneel to ask just one person to represent all of humanity).

17. E. O. Wilson (2012), p. 7, often speaks of our "Star Wars civilization": godlike technology, medieval institutions, and Stone Age emotions.

18. Asimov (1979 [1991]).

REFERENCES

Except where otherwise stated, all web links were operative as of October 30, 2014.

Abdul Latif Jameel Poverty Action Lab (JPAL). (n.d.). Encouraging teacher attendance through monitoring with cameras in rural Udaipur, India, www.povertyactionlab.org/evaluation/encouraging-teacher-attendance-through-monitoring-cameras-rural-udaipur-india.

Achebe, Chinua. (1977). An image of Africa: Racism in Conrad's "Heart of Darkness." *Massachusetts Review* 18(4):782–794.

———. (2011). Nigeria's promise, Africa's hope. *New York Times*, Jan. 16, 2011, www.nytimes.com/2011/01/16/opinion/16achebe.html.

Achenbach, Joel. (2014). Paul Farmer on Ebola: "This isn't a natural disaster, this is the terrorism of poverty." *Washington Post*, Oct. 6, 2014, www.washingtonpost.com/blogs/achenblog/wp/2014/10/06/paul-farmer-on-ebola-this-isnt-a-natural-disaster-this-is-the-terrorism-of-poverty/.

Agre, Philip E. (2002). Real-time politics: The Internet and the political process. *The Information Society* 18:311–331, www.tandfonline.com/doi/abs/10.1080/01972240290075174.

Agyare, Regina. (2014). Regina Agyare: Entrepreneur. LeanIn.org, http://leanin.org/stories/regina-agyare/.

Ainslie, George. (2001). *Breakdown of Will*. Cambridge University Press.

Al-Rasheed, Madawi. (2012). No Saudi spring: Anatomy of a failed revolution. *Boston Review*, March/April 2012, www.bostonreview.net/madawi-al-rasheed-arab-spring-saudi-arabia.

American Red Cross. (n.d.). Measles & Rubella Initiative, www.redcross.org/what-we-do/international-services/measles-initiative.

American Sociological Association. (2006). Peter H. Rossi (1921–2006). *Footnotes: Newsletter of the American Sociological Association*, Dec. 2006, 34(9), www.asanet.org/footnotes/dec06/indextwo.html.

Anderson, Chris. (2008). *The Long Tail: Why the Future of Business Is Selling Less of More*. Hyperion.

———. (2009). Q&A with Clay Shirky on Twitter and Iran. TED Blog, June 16, 2009, http://blog.ted.com/2009/06/16/qa_with_clay_sh/.

Angelucci, Manuela, Dean Karlan, and Jonathan Zinman. (2013). Win some lose some? Evidence from a randomized microcredit program placement experiment by Compartamos Banco. National Bureau of Economic Research Working Paper No. 19119, www.nber.org/papers/w19119.

Apostle, Hippocrates G. (1984). *Aristotle's Nicomachean Ethics*. Peripatetic Press.

Arora, Payal. (2010). Hope-in-the-Wall? A digital promise for free learning. *British Journal of Educational Technology* 41(5):689–702, http://onlinelibrary.wiley.com/doi/10.1111/j.1467-8535.2010.01078.x/abstract.

Arora, Payal, and Nimmi Rangaswamy, eds. (2014). ICTs for Leisure in Development Special Issue. *Information Technologies and International Development* 10(3), http://itidjournal.org/index.php/itid/issue/view/72.

Asimov, Isaac. (1942 [1991]). Runaround. In *Robot Visions*. Roc.

———. (1979 [1991]). The laws of robotics. In *Robot Visions*. Roc. Originally published in American Airlines in-flight magazine, 1979.

. (1985 [1994]). *Robots and Empire*. Collins.

———. (1986 [1994]). *Foundation and Earth*. Harper Collins.

Assmann, A. (1994). Wholesome knowledge: Concepts of wisdom in a historical and cross-cultural perspective. In D. L. Featherman, R. M. Learner, and M. Perlmutter, eds., *Life-Span Development and Behavior*. Lawrence Erlbaum, 12:187–224.

Atlantic. (2012). The conversation: Responses and reverberations. Aug. 22, 2012, www.theatlantic.com/magazine/archive/2012/09/the-conversation/309072/3/.

Auty, Richard M. (1993). *Sustaining Development in Mineral Economies: The Resource Curse Thesis*. Routledge.

Awuah, Patrick. (2012). Path to a new Africa. *Stanford Social Innovation Review*, Summer 2012, www.ssireview.org/articles/entry/path_to_a_new_africa.

Bajaj, Vikas. (2011). Microlenders, honored with Nobel, are struggling. *New York Times*, Jan. 5, 2011, www.nytimes.com/2011/01/06/business/global/06micro.html.

Bales, Kevin. (2002). The social psychology of modern slavery. *Scientific American* 286(4):80–88, www.scientificamerican.com/article/the-social-psychology-of/.

Banerjee, Abhijit, Shawn Cole, Esther Duflo, and Leigh Linden. (2007). Remedying education: Evidence from randomized experiments in India. *Quarterly Journal of Economics* 122(3):1235–1264, http://qje.oxfordjournals.org/content/122/3/1235.short.

Banerjee, Abhijit, and Esther Duflo. (2011a). *Poor Economics: A Radical Rethinking of the Way to Fight Global Poverty*. PublicAffairs.

Banerjee, Abhijit, Esther Duflo, Raghabendra Chattopadhyay, and Jeremy Shapiro. (2011b). Targeting the hard-core poor: An impact assessment, Nov. 2011, Jameel Poverty Action Lab, www.povertyactionlab.org/publication/targeting-hard-core-poor-impact-assessment.

Banerjee, Abhijit, Esther Duflo, Rachel Glennerster, and Cynthia Kinnan. (2010). The miracle of microfinance? Evidence from a randomized evaluation. Bread Working Paper No. 278, June 2010, http://ipl.econ.duke.edu/bread/system/files/bread_wpapers/379.pdf.

Bansal, Sarika. (2012). Shopping for a better world. *New York Times*, May 9, 2012, http://opinionator.blogs.nytimes.com/2012/05/09/shopping-for-a-better-world/.

Barcott, Rye. (2011). *It Happened on the Way to War: A Marine's Path to Peace*. Bloomsbury USA.

Barnes & Noble Booksellers. (n.d.). Barnes & Noble History, www.barnesandnobleinc.com/our_company/history/bn_history.html.

Barrera-Osorio, Felipe, and Leigh L. Linden. (2009). The use and misuse of computers in education: Evidence from a randomized experiment in Colombia. World Bank Policy Research Working Paper Series no. 4836, http://go.worldbank.org/8Q0F9DV040.

Bauerlein, Mark. (2009). *The Dumbest Generation: How the Digital Age Stupefies Young Americans and Jeopardizes Our Future*. Tarcher/Penguin.

Baumeister, Roy F., and Jessica L. Alquist. (2009). Is there a downside to good self-control? *Self and Identity* 8(2):115–130, www.tandfonline.com/doi/abs/10.1080/15298860802501474.

Baumeister, Roy F., and John Tierney. (2011). *Willpower: Rediscovering the Greatest Human Strength*. Penguin.

Baym, Nancy. (2010). *Personal Connections in the Digital Age*. Polity.

Behar, Anurag. (2010). Limits of ICT in education. *LiveMint*, Dec. 16, 2010, www.livemint.com/Opinion/Y3Rhb5CXMkGuUIyg4nrc3I/Limits-of-ICT-in-education.html.

———. (2012). Silver bullets in education. *LiveMint*, April 4, 2012, www.livemint.com/Opinion/BPS5faZixFQKrlUFGlivUL/Silver-bullets-in-education.html.

Bell, Daniel. (1999). *The Coming of Post-Industrial Society: A Venture in Social Forecasting*. Basic Books.

Bell, Genevieve. (2006). The age of the thumb: A cultural reading of mobile technologies from Asia. *Knowledge, Technology, & Policy*, Summer 2006,

19(2):41–57, http://link.springer.com/article/10.1007%2Fs12130-006-1023-5.

Bendavid, E., and J. Bhattacharya. (2009). The President's Emergency Plan for AIDS Relief in Africa: An evaluation of outcomes. *Annals of Internal Medicine* 150:688–695, http://annals.org/article.aspx?articleid=744499.

Benhabib, Jess, and Mark M. Spiegel. (1994). The role of human capital in economic development: Evidence from aggregate cross-country data. *Journal of Monetary Economics* 34(2):143–174, www.sciencedirect.com/science/article/pii/0304393294900477.

Bentham, Jeremy. (1789 [1907]). *An Introduction to the Principles of Morals and Legislation*. Library of Economics and Liberty, www.econlib.org/library/Bentham/bnthPML18.html.

Bentley, Daniel. (2012). First annual results of David Cameron's happiness index published. *The Independent*, July 24, 2012, www.independent.co.uk/news/uk/politics/first-annual-results-of-david-camerons-happiness-index-published-7972861.html.

Best, Michael L. (2004). Can the Internet be a human right? *Human Rights & Human Welfare* 4:23–31, https://www.du.edu/korbel/hrhw/volumes/2004/best-2004.pdf.

Bhatt, Kamla. (2003). IIT is an incredible institution, says Bill Gates. Rediff India Abroad, Jan. 18, 2003, www.rediff.com/money/2003/jan/18iit2.htm.

Bhide, Shashanka, and Aasha Kapu Mehta. (2004). Chronic poverty in rural India: Issues and findings from panel data. *Journal of Human Development* 5(2):195–209, www.tandfonline.com/doi/abs/10.1080/1464988042000225122.

Bilton, Nick. (2014). Steve Jobs was a low-tech parent. *New York Times*, Sept. 10, 2014, www.nytimes.com/2014/09/11/fashion/steve-jobs-apple-was-a-low-tech-parent.html.

Blackwood, Amy S., Katie L. Roeger, and Sarah L. Pettijohn. (2012). The nonprofit sector in brief: Public charities, giving, and volunteering, 2012, Urban Institute, www.urban.org/UploadedPDF/412674-The-Nonprofit-Sector-in-Brief.pdf.

Bloland, P., P. Simone, B. Burkholder, L. Slutsker, and K. M. De Cock. (2012). The role of public health institutions in global health system strengthening efforts: The US CDC's perspective. *PLOS Medicine* 9(4):e1001199, www.plosmedicine.org/article/info%3Adoi%2F10.1371%2Fjournal.pmed.1001199.

Bloomberg Businessweek. (2007). Compartamos: From non-profit to profit, Dec. 12, 2007, www.businessweek.com/stories/2007-12-12/compartamos-from-nonprofit-to-profit.

Boo, Katherine. (2012). *Behind the Beautiful Forevers: Life, Death, and Hope in a Mumbai Undercity*. Random House.

Bornstein, David. (2004). *How to Change the World: Social Entrepreneurs and the Power of New Ideas*. Oxford University Press.

Bornstein, David. (2014). Where YouTube meets the farm. *New York Times*, April 3, 2014, http://opinionator.blogs.nytimes.com/2013/04/03/where-youtube-meets-the-farm/.

Boston Review. (2010). Ideas matter: Can technology solve global poverty? Video available as item #38 at https://itunes.apple.com/us/itunes-u/media/id433442313.

Boudreaux, Donald J., and Mark J. Perry. (2013). Donald Boudreaux and Mark Perry: The myth of a stagnant middle class. *Wall Street Journal*, Jan. 23, 2013, http://online.wsj.com/news/articles/SB10001424127887323468604578249723138161566.

Bourdieu, Pierre. (1979 [1984]). *Distinction: A Social Critique of the Judgement of Taste*. Richard Nice, trans. Harvard University Press.

Brasor, Philip. (2001). "Comfort" education at expense of standards? *Japan Times*, Sept. 23, 2001, www.japantimes.co.jp/text/fl20010923pb.html.

Brewis, Alexandra A., Amber Wutich, Ashlan Falletta-Cowden, and Isa Rodriguez-Soto. (2011). Body norms and fat stigma in global perspective. *Current Anthropology* 52(2):269–276, www.jstor.org/stable/10.1086/659309.

Brill, Steven. (2013). Bitter pill: How outrageous pricing and egregious profits are destroying our health care. *Time*, March 4, 2013, 16–55, http://time.com/198/bitter-pill-why-medical-bills-are-killing-us/.

Brounstein, Marty. (2000). *Coaching & Mentoring for Dummies*. Wiley.

Bueno de Mesquita, Bruce, and Alastair Smith. (2011). *The Dictator's Handbook: Why Bad Behavior Is Almost Always Good Politics*. PublicAffairs.

Butler, Kiera. (2014). Do Toms Shoes really help people? *Mother Jones*, May 14, 2014, www.motherjones.com/environment/2012/05/toms-shoes-buy-one-give-one.

Butterfield, L. H., Marc Friedlaender, and Mary-Jo Kline, eds. (1975). *The Book of Abigail and John: Selected Letters of the Adams Family, 1762–1784*. Harvard University Press.

Cairncross, Frances. (1997). *The Death of Distance: How the Communications Revolution Will Change Our Lives*. Harvard Business School Press.

———. (2001). *The Death of Distance 2.0: How the Communications Revolution Will Change Our Lives*. Texere.

Camerer, Colin F., and Richard H. Thaler. (1995). Anomalies: Ultimatums, dictators and manners. *Journal of Economic Perspectives* 9(2):209–219, www.aeaweb.org/articles.php?doi=10.1257/jep.9.2.209.

Cameron, William Bruce. (1963). *Informal Sociology: A Casual Introduction to Sociological Thinking*. Random House.

Caplan, Bryan. (2012). Selfish reasons to have more kids: Why being a great parent is less work and more fun than you think. Basic Books.

Carlin, George. (1984). Carlin on campus. *HBO*, April 19, 1984.

Carolina for Kibera. (n.d.). Stories: Steve Juma, http://cfk.unc.edu/ourimpact /stories/.

Carr, Nicholas. (2011). *The Shallows: What the Internet Is Doing to Our Brains*. W. W. Norton.

CBS News. (2007). What if every child had a laptop? May 20, 2007, www.cbsnews.com/news/what-if-every-child-had-a-laptop/.

Center for American Women and Politics. (2014). 2014: Not a landmark year for women, despite some notable firsts. *CAWP Election Watch*, www.cawp.rutgers .edu/press_room/news/documents/PressRelease_11-05-14-electionresults.pdf.

Centers for Disease Control and Prevention. (2008). Progress toward strengthening blood transfusion services – 14 countries, 2003–2007. *Morbidity and Mortality Weekly Report* 57(47):1273–1277, www.cdc.gov /mmwr/preview/mmwrhtml/mm5747a2.htm.

Chang, Ha-Joon. (2010). *23 Things They Don't Tell You About Capitalism*. Bloomsbury.

Chew, Han-Ei, Mark Levy, and Vigneswaran Ilavarasan. (2011). The limited impact of ICTs on microenterprise growth: A study of businesses owned by women in urban India. *Information Technologies and International Development* 7(4):1–16, http://itidjournal.org/itid/article/view/788.

Chulov, Martin. (2012). Syria shuts off Internet access across the country. *The Guardian*, Nov. 29, 2012, www.theguardian.com/world/2012/nov/29/syria -blocks-internet.

Clinton, Hillary. (2010). Remarks on Internet freedom. The Newseum, Jan. 21, 2010, www.state.gov/secretary/20092013clinton/rm/2010/01/135519.htm.

CNN. (2011). Ghonim: Facebook to thank for freedom. Feb. 11, 2011, http://edition.cnn.com/video/#/video/bestoftv/2011/02/11/exp.ghonim .facebook.thanks.cnn.

Cobb, Clifford, Ted Halstead, and Jonathan Rowe. (1995). If the GDP is up, why is America down? *The Atlantic*, Oct. 1995, www.theatlantic.com/past /politics/ecbig/gdp.htm.

Cohen, Roger. (2011). Revolutionary Arab geeks. *New York Times*, Jan. 27, 2011, www.nytimes.com/2011/01/28/opinion/28iht-edcohen28.html.

Cohen, W., and D. Levinthal. (1990). Absorptive capacity: A new perspective on learning and innovation. *Administrative Science Quarterly* 35(1):128–152, www.jstor.org/stable/2393553.

Cohn, Jonathan. (2008). *Sick: The Untold Story of America's Health Care Crisis – and the People Who Pay the Price*. Harper Perennial.

Coleman, James S., et al. (1966). Equality of educational opportunity. US Department of Health, Education, and Welfare. US Government Printing Office.

Coleman-Jensen, Alisha, Mark Nord, and Anita Singh. (2013). Household food security in the United States 2012. US Department of Agriculture. Economic Research Report No. 155, Sept. 2013, www.ers.usda.gov/publications/err-economic-research-report/err155.aspx.

Colindres, P., J. Mermin, E. Ezati, S. Kambabazi, P. Buyungo, L. Sekabembe, F. Baryarama, F. Kitabire, S. Mukasa, F. Kizito, C. Fitzgerald, and R. Quick. (2008). Utilization of a basic care and prevention package by HIV-infected persons in Uganda. *AIDS Care* 20(2):139–145, www.tandfonline.com/doi/abs/10.1080/09540120701506804.

Collins, Darryl, Jonathan Morduch, Stuart Rutherford, and Orlanda Ruthven. (2009). *Portfolios of the Poor: How the World's Poor Live on $2 a Day*. Princeton University Press.

Constitution of India. (2011). Government of India, http://india.gov.in/my-government/constitution-india/constitution-india-full-text.

Consultative Group to Assist the Poor (CGAP). (2008). Are we overestimating demand for microloans? *CGAP Brief*, April 2008, www.cgap.org/sites/default/files/CGAP-Brief-Are-We-Overestimating-Demand-for-Microloans-Apr-2008.pdf.

Coontz, Stephanie. (2014). The new instability. *New York Times*, July 26, 2014, www.nytimes.com/2014/07/27/opinion/sunday/the-new-instability.html.

Corelli, Marie. (1905). The spirit of work. In The Daily Mail, ed., *The Queen's Carol: An Anthology of Poems, Stories, Essays, Drawings, and Music*. The Daily Mail.

Corporation for National and Community Service. (2006). Volunteer growth in America: A review of trends since 1974. Dec. 2006, www.nationalservice.gov/pdf/06_1203_volunteer_growth.pdf.

Counts, Alex. (2008). *Small Loans, Big Dreams: How Nobel Prize Winner Muhammad Yunus and Microfinance Are Changing the World*. John Wiley and Sons.

Crescenzi, Riccardo, Max Nathan, and Andrés Rodríguez-Pose. (2013). Do inventors talk to strangers? On proximity and collaborative knowledge creation. IZA Discussion Paper No. 7797, http://ssrn.com/abstract=2367672.

Criminisi, Antonio, Patrick Pérez, and Kentaro Toyama. (2004). Region filling and object removal by exemplar-based image inpainting. *IEEE Transactions on Image Processing* 13(9):1200–1212, http://dx.doi.org/10.1109/TIP.2004.833105.

Cristia, Julián P., Pablo Ibarrarán, Santiago Cueto, Ana Santiago, and Eugenio Severín. (2012). Technology and child development: Evidence from the One Laptop Per Child Program. IDB Working Paper Series, No. IDB-WP-304, http://idbdocs.iadb.org/wsdocs/getdocument.aspx?docnum=36706954.

CTIA. (2011). More wireless devices than Americans. CTIA Blog, Oct. 11, 2011, http://blog.ctia.org/2011/10/11/more-wireless-devices-than-americans/.

Cuban, Larry. (1986). *Teachers and Machines: The Classroom Use of Technology Since 1920*. Teachers College Press.

Dahl, Fredrik. (2010). Iran's police vow no tolerance towards protesters. Reuters, Feb. 6, 2010, www.reuters.com/article/2010/02/06/idUSDAH650941.

Dahl, Robert Alan. (1971). *Polyarchy: Participation and Opposition*. Yale University Press.

Darrow, Benjamin. (1932). *Radio: The Assistant Teacher*. R. G. Adams.

Davidson, Richard. (1992). Emotion and affective style: Hemispheric substrates. *Psychological Science* 3:39–43, http://pss.sagepub.com/content/3/1/39.abstract.

Davidson, R., D. Jackson, and N. Kalin. (2000). Emotion, plasticity, context, and regulation: Perspectives from affective neuroscience. *Psychological Bulletin* 126:890–906, http://psycnet.apa.org/journals/bul/126/6/890/.

Davis, Kristin, Ephraim Nkonya, Edward Kato, Daniel Ayalew Mekonnen, Martins Odendo, Richard Miiro, and Jackson Nkurba. (2010). Impact of Farmer Field Schools on agricultural productivity and poverty in East Africa. IFPRI Discussion Paper No. 00992. International Food Policy and Research Institute, www.ifpri.org/sites/default/files/publications/ifpridp00992.pdf.

Deci, E. L., and R. M. Ryan. (1985). *Intrinsic Motivation and Self-Determination in Human Behavior*. Plenum.

———. (1991). A motivational approach to self: Integration in personality. Pp. 237–288 in *Nebraska Symposium on Motivation*, vol. 38, *Perspectives on Motivation*. University of Nebraska Press.

Deininger, Klaus, and John Okidi. (2003). Growth and poverty reduction in Uganda, 1999–2000: Panel data evidence. *Development Policy Review* 21:481–509, July 2003, http://onlinelibrary.wiley.com/doi/10.1111/1467-7679.00220/abstract.

Delforge, Pierre. (2014). America's data centers consuming and wasting growing amounts of energy. Natural Resources Defense Council, Dec. 15, 2014, www.nrdc.org/energy/data-center-efficiency-assessment.asp.

Dell, Nicola, Vidya Vaidyanathan, Indrani Medhi, Edward Cutrell, and William Thies. (2012). "Yours is better!": Participant response bias in HCI. Pp. 1321–1330 in *Proceedings of the SIGCHI Conference on Human Factors in Computing Systems (CHI '12)*. ACM, http://doi.acm.org/10.1145/2207676.2208589.

De Melo, Gioia, Alina Machado, Alfonso Miranda, and Magdalena Viera. (2014). Impacto del Plan Ceibal en el aprendizaje: Evidencia de la mayor experiencia OLPC. Instituto de Economia, www.iecon.ccee.edu.uy/dt-13-13-impacto-del-plan-ceibal-en-el-aprendizaje-evidencia-de-la-mayor-experiencia-olpc/publicacion/376/es/.

De Munck, Bert, Steven L. Kaplan, and Hugo Soly. (2007). *Learning on the Shop Floor: Historical Perspectives on Apprenticeship*. Berghahn.

DeNavas-Walt, Carmen, Bernadette D. Proctor, and Jessica C. Smith. (2009). US Census Bureau, Current Population Reports, P60-236, Income, Poverty, and Health Insurance Coverage in the United States: 2008, US Government Printing Office, www.census.gov/prod/2009pubs/p60-236.pdf.

DeRenzi, Brian, Leah Findlater, Jonathan Payne, Benjamin Birnbaum, Joachim Mangilima, Tapan Parikh, Gaetano Borriello, and Neal Lesh. (2012). Improving community health worker performance through automated SMS. Pp. 25–34 in *Proceedings of the Fifth International Conference on Information and Communication Technologies and Development*, http://doi.acm.org/10.1145/2160673.2160677.

Deutsch, David. (2011). *The Beginning of Infinity: Explanations That Transform the World*. Viking.

Development Initiatives. (2010). GHA report 2010, www.globalhumanitarianassistance.org/wp-content/uploads/2010/07/GHA_Report8.pdf.

Diamandis, Peter H., and Steven Kotler. (2012). *Abundance: The Future Is Better Than You Think*. Free Press.

Diamond, Jared. (2008). What's your consumption factor? *New York Times*, Jan. 2, 2008, www.nytimes.com/2008/01/02/opinion/02diamond.html.

Diener, Ed, and Robert Biswas-Diener. (2008). *Happiness: Unlocking the Mysteries of Psychological Wealth*. Blackwell.

Diener, Ed, Eunkook M. Suh, Richard E. Lucas, and Heidi L. Smith. (1999). Subjective well-being: Three decades of progress. *Psychological Bulletin* 125(2):276–302, http://psycnet.apa.org/doi/10.1037/0033-2909.125.2.276.

Digital StudyHall. (n.d.). Digital StudyHall website, http://dsh.cs.washington.edu.

Dimitrov, Martin. (2008). The resilient authoritarians. *Current History* 107(705):24–29, www.currenthistory.com/Article.php?ID=514.

Drexler, Alejandro, Greg Fischer, and Antoinette Schoar. (2010). Keeping it simple: Financial literacy and rules of thumb, Sept. 2010. CEPR Discussion Paper No. DP7994, http://ssrn.com/abstract=1707884.

Driver, Julia. (2001). *Uneasy Virtue*. Cambridge Studies in Philosophy. Cambridge University Press.

Dudley, Brier. (2009). Microsoft alum's university taking root. *Seattle Times*, Feb. 2, 2009, http://seattletimes.com/html/businesstechnology/2008696526_brier02.html.

Duflo, Esther. (2012). Human values and the design of the fight against poverty. Tanner Lectures, May 2–4, 2012, www.povertyactionlab.org/doc/esther-duflos-tanner-lecture.

Duflo, Esther, Rema Hanna, and Stephen P. Ryan. (2012). Incentives work: Getting teachers to come to school. *American Economic Review* 102(4):1241–1278, https://www.aeaweb.org/articles.php?doi=10.1257/aer.102.4.1241.

Du Gay, Paul. (1997). *Doing Cultural Studies: The Story of the Sony Walkman*. Bath Press Colourbooks, Open University.

Duncan, Arne. (2012). The new platform for learning. Keynote delivered at South by Southwest conference, www.ed.gov/news/speeches/new-platform-learning.

Duthiers, Vladimir, and Jessica Ellis. (2013). Millionaire who quit Microsoft to educate Africa's future leaders. CNN, May 1, 2013, www.cnn.com/2013/05/01/world/africa/patrick-awuah-ashesi-ghana/index.html.

Dweck, Carol. (2007). *Mindset: The New Psychology of Success*. Ballantine.

Eagan, K., J. B. Lozano, S. Hurtado, and M. H. Case. (2013). The American freshman: National norms fall 2013. Los Angeles: Higher Education Research Institute, UCLA, www.heri.ucla.edu/monographs/TheAmericanFreshman2013.pdf.

Easterly, William. (2001). *The Elusive Quest for Growth: Economists' Adventures and Misadventures in the Tropics*. MIT Press.

———. (2006). *The White Man's Burden: Why the West's Efforts to Aid the Rest Have Done So Much Ill and So Little Good*. Penguin.

———. (2014). *The Tyranny of Experts: Economists, Dictators, and the Forgotten Rights of the Poor*. Basic Books.

Easterly, William, and Laura Freschi. (2012). Save the poor Beltway bandits! NYU Development Research Institute Blog, May 7, 2012, http://nyudri.org/2012/05/07/save-the-poor-beltway-bandits/.

Economist. (1977). The Dutch disease. Nov. 26, 1977, 82–83.

———. (2005). The real digital divide. May 10, 2005, www.economist.com/node/3742817.

———. (2008). Malthus, the false prophet. May 15, 2008, www.economist.com/node/11374623.

———. (2014). Cutting down on cutting down. June 7, 2014, www.economist.com/news/science-and-technology/21603409-how-brazil-became-world-leader-reducing-environmental-degradation-cutting.

Ehrenreich, Barbara. (2009). *Bright-Sided: How the Relentless Promotion of Positive Thinking Has Undermined America*. Metropolitan Books.

Eigsti, Inge-Marie, Vivian Zayas, Walter Mischel, Yuichi Shoda, Ozlem Ayduk, Mamta B. Dadlani, Matthew C. Davidson, J. Lawrence Aber,

and B. J. Casey. (2006). Predicting cognitive control from preschool to late adolescence and young adulthood. *Psychological Science* 17(6):478–484, http://pss.sagepub.com/content/17/6/478.abstract.

Ellerman, David. (2005). *Helping People Help Themselves: From the World Bank to an Alternative Philosophy of Development Assistance*. University of Michigan Press.

Ellul, Jacques. (1964). *The Technological Society*. John Wilkinson, trans. Vintage.

———. (1965 [1973]). *Propaganda: The Formation of Men's Attitudes*. Konrad Kellen and Jean Lerner, trans. Vintage.

El Sistema. (n.d.). Lista de nucleos por estado, http://fundamusical.org.ve/nucleos/.

Engels, Friedrich. (1844 [1968]). *The Condition of the Working Class in England*. W. O. Henderson and W. H. Chaloner, trans. Stanford University Press.

Ericsson. (2014). Ericsson mobility report: On the pulse of the networked society, www.ericsson.com/res/docs/2014/ericsson-mobility-report-june-2014.pdf.

Erikson, Erik H. (1950). *Childhood and Society*. Norton.

Etzioni, Amitai. (1993). *The Spirit of Community: Rights, Responsibilities, and the Communitarian Agenda*. Crown.

Evans, Peter. (2005). The challenges of the institutional turn: New interdisciplinary opportunities in development theory. Pp. 90–116 in Victor Nee and Richard Sweberg, eds., *The Economic Sociology of Capitalist Institutions*. Princeton University Press.

Fairlie, Robert W., and Jonathan Robinson. (2013). Experimental evidence on the effects of home computers on academic achievement among schoolchildren. *American Economic Journal: Applied Economics* 5(3):211–240, https://www.aeaweb.org/articles.php?doi=10.1257/app.5.3.211.

Farmer, Paul. (2005). *Pathologies of Power: Health, Human Rights, and the New War on the Poor*. University of California Press.

Feenberg, Andrew. (1999). *Questioning Technology*. Routledge.

Festinger, Leon. (1957). *A Theory of Cognitive Dissonance*. Stanford University Press.

Findlater, Leah, Ravin Balakrishnan, and Kentaro Toyama. (2009). Comparing semiliterate and illiterate users' ability to transition from audio+text to text-only interaction. Pp. 1751–1760 in *Proceedings of the SIGCHI Conference on Human Factors in Computing Systems (CHI '09)*. ACM, http://doi.acm.org/10.1145/1518701.1518971.

Fisher, Lawrence M. (1988). Moving up fast in the software sweepstakes. *New York Times*, Feb. 28, 1988, www.nytimes.com/1988/02/28/business/moving-up-fast-in-the-software-sweepstakes.html.

Fisher, Max. (2012). Here's a map of the countries that provide universal health care (America's still not on it). *The Atlantic*, June 28, 2012, www.theatlantic.com/international/archive/2012/06/heres-a-map-of-the-countries-that-provide-universal-health-care-americas-still-not-on-it/259153/.

Fiske, Alan Page. (1993). *Structures of Social Life: The Four Elementary Forms of Human Relations*. Free Press.

Florida, Richard. (2002). *The Rise of the Creative Class*. Basic Books.

Food and Agriculture Organization. (2013). Global hunger down but millions still chronically hungry. News release, Oct. 1, 2013, www.fao.org/news/story/en/item/198105/icode/.

Fowler, James W. (1981). *The Stages of Faith: The Psychology of Human Development and the Quest for Meaning*. HarperCollins.

Frank, Robert H. (2012). Will the skillful win? They should be so lucky. *New York Times*, Aug. 5, 2012, www.nytimes.com/2012/08/05/business/of-luck-and-success-economic-view.html.

Frank, Robert H., and Philip J. Cook. (1996). *The Winner-Take-All Society: Why the Few at the Top Get So Much More Than the Rest of Us*. Penguin.

Franklin, Benjamin. (1986). *The Autobiography and Other Writings*. Viking Penguin.

Franzen, Jonathan. (2010). *The Corrections*. Farrar, Straus and Giroux.

Fraser, Barbara. (2012). "Improved" cookstoves may do little to reduce harmful indoor emissions. *Scientific American*, July 11, 2012, www.scientificamerican.com/article/improved-cookstoves-little-reduce-harmful-indoor-emissions/.

Freud, Sigmund. (1962). *Three Essays on the Theory of Sexuality*. James Strachey, trans. Basic Books.

Fried, Barbara. (2013). Beyond blame. *Boston Review*, June 28, 2013, www.bostonreview.net/forum/barbara-fried-beyond-blame-moral-responsibility-philosophy-law.

Friedman, Thomas L. (2006). *The World Is Flat: The Globalized World in the Twenty-First Century*. Penguin.

Fukuyama, Francis. (1992). *The End of History and the Last Man*. Free Press.

Fuller, Robert W. (2004). *Somebodies and Nobodies: Overcoming the Abuse of Rank*. New Society Publishers.

Fundación Paraguaya. (n.d.). Self-sufficient school, www.fundacionparaguaya.org.py/?page_id=741.

Gallup and Purdue University. (2014). Great jobs great lives: The 2014 Gallup-Purdue index report, http://products.gallup.com/168857/gallup-purdue-index-inaugural-national-report.aspx.

Gander, Kashmira. (2014). Kenyan Cardinal John Njue tells congregation that tetanus vaccination programme for pregnant women "is a bit fishy." *The Independent*, March 25, 2014, www.independent.co.uk/news/world/africa/kenyan-cardinal-john-njue-tells-congregation-that-tetanus-vaccination-programme-for-pregnant-women-is-a-bit-fishy-9214994.html.

Gandhi, Rikin, Rajesh Veeraraghavan, Kentaro Toyama, and Vanaja Ramprasad. (2009). Digital Green: Participatory video for agricultural extension. *Information Technologies and International Development* 5(1):1–15, http://itidjournal.org/itid/article/view/322/145.

Gates, Bill. (1995). *The Road Ahead*. Viking.

Geertz, Clifford. (1984). Anti anti-relativism. *American Anthropologist* 86(2):263–278, www.jstor.org/stable/678960.

George, Abraham. (2004). *India Untouched: The Forgotten Face of Rural Poverty*. East West Books.

Gilbert, Daniel. (2006). *Stumbling on Happiness*. Vintage.

Giving USA. (2014). Giving USA 2014 highlights, http://store.givingusareports.org/Giving-USA-2014-Report-Highlights-P114.aspx.

Gladwell, Malcolm. (2011). Does Egypt need Twitter? *New Yorker* News Desk, Feb. 2, 2011, www.newyorker.com/online/blogs/newsdesk/2011/02/does-egypt-need-twitter.html.

Goldin, Claudia, and Lawrence Katz. (2009). *The Race Between Education and Technology*. Harvard University Press.

Goldman, David. (2009). Obama's big idea: Digital health records. CNNMoney.com, Jan. 12, 2009, http://money.cnn.com/2009/01/12/technology/stimulus_health_care/.

Goleman, Daniel. (1990). Probing school success of Asian Americans. *New York Times*, Sept. 11, 1990, www.nytimes.com/1990/09/11/science/probing-school-success-of-asian-americans.html.

———. (1995). *Emotional Intelligence*. Bantam Books.

GOP Doctors Caucus. (n.d.). Health information technology, http://doctorscaucus.gingrey.house.gov/issues/issue/?IssueID=9947.

Grameen Foundation. (2014). Lessons learned, 2009–2014: Community Knowledge Worker Uganda Program. Oct. 2014, http://grameenfoundation.org/sites/grameenfoundation.org/files/resources/Grameen-Foundation_CKW-Lessons-Learned-%282009-2014%29_Executive-Summary_0.pdf.

Green, Elizabeth. (2014). Why do Americans stink at math? *New York Times Magazine*, July 23, 2014, www.nytimes.com/2014/07/27/magazine/why-do-americans-stink-at-math.html.

Grenfell, Michael, ed. (2008). *Pierre Bourdieu: Key Concepts*. Acumen.

Guilford, Gwynn. (2013). In China, being retweeted 500 times can land you in jail. *The Atlantic*, Sept. 20, 2013, www.theatlantic.com/china/archive/2013/09/in-china-being-retweeted-500-times-can-land-you-in-jail/279859/.

Hacker, Jacob S., and Paul Pierson. (2010). *Winner-Take-All Politics: How Washington Made the Rich Richer—and Turned Its Back on the Middle Class*. Simon and Schuster.

Hafner, Katie. (2009). Texting may be taking a toll. *New York Times*, May 25, 2009, www.nytimes.com/2009/05/26/health/26teen.html.

Haggbloom, Steven J., Renee Warnick, Jason E. Warnick, Vinessa K. Jones, Gary L. Yarbrough, Tenea M. Russell, Chris M. Borecky, Reagan McGahhey, John L. Powell III, Jamie Beavers, and Emmanuelle Monte. (2002). The 100 most eminent psychologists of the 20th century. *Review of General Psychology* 6(2):139–152, http://psycnet.apa.org/journals/gpr/6/2/139/.

Haidt, Jonathan. (2006). *The Happiness Hypothesis: Finding Modern Truth in Ancient Wisdom*. Basic Books.

Hampton, Keith N., Lauren Sessions Goulet, Lee Rainie, and Kristen Purcell. (2011). Social networking sites and our lives: How people's trust, personal relationships, and civic and political involvement are connected to their use of social networking sites and other technologies. Pew Research Center, June 16, 2011, http://pewinternet.org/Reports/2011/Technology-and-social-networks.aspx.

Harris, Judith Rich. (2009). *The Nurture Assumption: Why Children Turn Out the Way They Do*, rev ed. Free Press.

Harvey, T. A. (1988). How Sony Corp. became first with kids. *AdWeek's Marketing Week*, Nov. 21, 1988, 58–59.

Hauslohner, Abigail. (2011). Is Egypt about to have a Facebook Revolution? *Time*, Jan. 24, 2011, www.time.com/time/world/article/0,8599,2044142,00.html.

Hazell, Peter B. R. (2009). The Asian Green Revolution. IFPRI Discussion Paper. International Food Policy Research Institute, www.ifpri.org/sites/default/files/publications/ifpridp00911.pdf.

Heckman, James J. (1997). Instrumental variables: A study of implicit behavioral assumptions used in making program evaluations. *Journal of Human Resources* 32(3):441–462, www.jstor.org/stable/pdfplus/146178.

———. (2012). Promoting social mobility. *Boston Review*, Sept./Oct. 2012, www.bostonreview.net/BR37.5/ndf_james_heckman_social_mobility.php.

Heeks, Richard. (2009). Worldwide expenditures on ICT4D. ICTs for Development blog, https://ict4dblog.wordpress.com/2009/04/06/world wide-expenditure-on-ict4d.

Hegewisch, Ariane, and Claudia Williams. (2013). The gender wage gap: 2012. Institute for Women's Policy Research, Fact Sheet IWPR #C350,www.iwpr .org/publications/pubs/the-gender-wage-gap-2012-1/.

Heilbroner, Robert L. (1967). Do machines make history? *Technology and Culture* 8:335–345.

———. (1994). Technological determinism revisited. Pp. 67–78 in Merritt Roe Smith and Leo Marx, eds., *Does Technology Drive History? The Dilemma of Technological Determinism*. MIT Press.

Henrich, Joseph, Robert Boyd, Samuel Bowles, Colin Camerer, Ernst Fehr, and Herbert Gintis. (2004). *Foundations of Human Sociality: Economic Experiments and Ethnographic Evidence from Fifteen Small-Scale Societies*. Oxford University Press.

Henrich, Joseph, Steven J. Heine, and Ara Norenzayan. (2010). The WEIRDest people in the world? *Behavioral and Brain Sciences* 33(2–3):61–83, http://psycnet.apa.org/doi/10.1017/S0140525X0999152X.

Higher Education Research Institute. (2008). The American freshman: 1966–2006. HERI Research Brief, Jan. 2008, http://heri.ucla.edu/PDFs/pubs /briefs/40yrTrendsResearchBrief.pdf.

Hindu. (2006). Nod for massive e-governance scheme. Sept. 22, 2006, www.thehindu.com/todays-paper/tp-national/nod-for-massive-egovernance -scheme/article3079295.ece.

———. (2014). PM hails Mars mission as "historic." Sept. 24, 2014, www.thehindu.com/news/national/pm-hails-mars-mission-success-as-historic /article6441323.ece.

Hoffman, Edward, ed. (1996). *Future Visions: The Unpublished Papers of Abraham Maslow*. Sage.

Hoffman, Jan. (2010). On top of the happiness racket. *New York Times*, Feb. 28, 2010, www.nytimes.com/2010/02/28/fashion/28rubin.html.

Hofstede, Geert. (1984). The cultural relativity of the quality of life concept. *Academy of Management Review* 9(3):389–398, http://amr.aom.org /content/9/3/389.abstract?related-urls=yes&legid=amr;9/3/389.

Hu, Ruifa, Jikun Huang, and Kevin Z. Chen. (2012). The public agriculture extension system in China: Development and reform. Roundtable Consultation on Agricultural Extension, Beijing, March 15–17, 2012, www.syngentafoundation.org/__temp/HU_HUANG_CHEN_AG_EXTN_CHINA_DEVELOPMENT_REFORM.pdf.

Hu, Winnie. (2007). Seeing no progress, some schools drop laptops. *New York Times*, May 4, 2007, www.nytimes.com/2007/05/04/education/04laptop.html.

Hume, David. (1740 [2011]). A treatise of human nature: Being an attempt to introduce the experimental method of reasoning into moral subjects. In *Hume: The Essential Philosophical Works*. Wordsworth Editions Limited.

Hutchful, David, Akhil Mathur, Apurva Joshi, and Edward Cutrell. (2010). Cloze: An authoring tool for teachers with low computer proficiency. In *Proceedings of the International Conference on Information and Communication Technologies and Development (ICTD2010)*. London, 2010, http://dl.acm.org/citation.cfm?id=2369239.

Huxley, Aldous. (2005). *Brave New World and Brave New World Revisited*. Modern Classics ed. Harper Perennial.

IBNLive. (2010). Karnataka CM wants me to go: Lokayukta. June 25, 2010, http://ibnlive.in.com/news/karnataka-cm-wants-me-to-go-lokayukta/125308-37-67.html.

Indian Institute for Human Settlements. (2012). Urban India 2011: Evidence. 2011 India Urban Conference, http://iihs.co.in/wp-content/uploads/2013/12/IUC-Book.pdf.

Indian Space Research Organization. (2008). Press Release: Chandrayaan-1 successfully enters lunar orbit, Nov. 8, 2008, www.isro.org/pressrelease/scripts/pressreleasein.aspx?Nov08_2008.

Inglehart, Ronald, Roberto Foa, Christopher Peterson, and Christian Welzel. (2008). Development freedom, and rising happiness: A global perspective (1981–2007). *Perspectives on Psychological Science* 3(4):264–285, http://pps.sagepub.com/content/3/4/264.abstract.

Inglehart, Ronald, and Pippa Norris. (2003). *Rising Tide: Gender Equality and Cultural Change Around the World*. Cambridge University Press.

Inglehart, Ronald, and Christian Welzel. (2005). *Modernization, Cultural Change and Democracy*. Cambridge University Press.

———. (2010). Changing mass priorities: The link between modernization and democracy. *Perspectives on Politics* 8(2):554, www.worldvaluessurvey.org/wvs/articles/folder_published/article_base_54.

International Coach Federation. (n.d.). FAQs, www.coachfederation.org/about/landing.cfm?ItemNumber=844&navItemNumber=617.

International Committee of the Red Cross. (2014). Annual report 2013, www.icrc.org/eng/assets/files/annual-report/icrc-annual-report-2013.pdf.

International Math Olympiad. (2014). Results, www.imo-official.org/results.aspx.

International Telecommunications Union (ITU). (2014). ITU releases 2014 ICT figures, www.itu.int/net/pressoffice/press_releases/2014/23.aspx.

Internet.org. (2013). Technology leaders launch partnership to make Internet access available to all. Press release, Aug. 20, 2013, https://fbcdn-dragon-a.akamaihd.net/hphotos-ak-prn1/851572_595418650524294_1430606495_n.pdf.

Isen, Alice M., and Paula F. Levin. (1972). Effect of feeling good on helping: Cookies and kindness. *Journal of Personality and Social Psychology* 21(3):384–388, http://psycnet.apa.org/doi/10.1037/h0032317.

I-TECH. (2011). About I-TECH, www.go2itech.org/who-we-are/about-i-tech.

———. (2013). The leader who says "I can." Everyday Leadership.org, www.everydayleadership.org/video/p519.

Jack, William, and Tavneet Suri. (2011). Mobile money: The economics of M-PESA. National Bureau of Economic Research Working Paper 16721, www.nber.org/papers/w16721.pdf.

Jakiela, Pamela, Edward Miguel, and Vera te Velde. (2012). You've earned it: Combining field and lab experiments to estimate the impact of human capital on social preferences. National Bureau of Economic Research Working Paper 16449, https://ideas.repec.org/p/nbr/nberwo/16449.html.

Jasanoff, Sheila. (2002). New modernities: Reimagining science, technology, and development. *Environmental Values* 11:253–276, www.ingentaconnect.com/content/whp/ev/2002/00000011/00000003/art00001.

Jensen, Derrick. (1998 [2005]). Actions speak louder than words. In John Zerzan, ed., *Against Civilization: Readings and Reflections*. Feral House.

Jensen, Robert. (2012). Do labor market opportunities affect young women's work and family decisions? Experimental evidence from India. *Quarterly Journal of Economics* 127(2):753–792, http://qje.oxfordjournals.org/content/early/2012/03/02/qje.qjs002.abstract.

Jhunjhunwala, Ashok, Anuradha Ramachandran, and Alankar Bandyopadhyay. (2004). n-Logue: The story of a rural service provider in India. *Journal of Community Informatics* 1(1):30–38, http://unpan1.un.org/intradoc/groups/public/documents/APCITY/UNPAN023008.pdf.

Johnson, W. Brad, and Charles R. Ridley. (2008). *The Elements of Mentoring*, rev. ed. Palgrave Macmillan.

Jones, Michael L. (2006). The growth of nonprofits. *Bridgewater Review* 25(1):13–17, http://vc.bridgew.edu/br_rev/vol25/iss1/8.

Karlan, Dean, and Jonathan Zinman. (2010). Expanding credit access: Using randomized supply decisions to estimate the impacts. *Review of Financial Studies* 23:433–464, http://rfs.oxfordjournals.org/content/23/1/433.

———. (2011). Microcredit in theory and practice: Using randomized credit scoring for impact evaluation. *Science* 332(6035):1278–1284, www.sciencemag.org/content/332/6035/1278.long.

Karnani, Aneel. (2007). The mirage of marketing to the bottom of the pyramid: How the private sector can help alleviate poverty. *California Management Review* 49(4):90–111, www.jstor.org/stable/41166407.

Kato, Mariko. (2009). Education chief takes liberal path. *Japan Times*, Oct. 9, 2009, www.japantimes.co.jp/text/nn20091009f2.html.

Kekic, Laza. (2007). The Economist Intelligence Unit's index of democracy. Economist Intelligence Unit, www.economist.com/media/pdf/DEMOCRACY_INDEX_2007_v3.pdf.

Khan, Ismail. (2013). Anti-polio campaign worker is shot dead in Pakistan. *New York Times*, May 28, 2013, www.nytimes.com/2013/05/29/world/asia/anti-polio-campaign-worker-shot-dead-in-pakistan.html.

Kidder, Tracy. (2009). *Strength in What Remains*. Random House.

King, Gary, Jennifer Pan, and Margaret E. Roberts. (2013a). How censorship in China allows government criticism but silences collective expression. *American Political Science Review* 107(2):1–18, http://dx.doi.org/10.1017/S0003055413000014.

———. (2013b). Randomized experimentation and participant observation. *Science* 345(6199):1–10, www.sciencemag.org/content/345/6199/1251722.

Kirkpatrick, David D., and Mona El-Naggar. (2011). Qaddafi's grip falters as his forces take on protesters. *New York Times*, Feb. 22, 2011, www.nytimes.com/2011/02/22/world/africa/22libya.html.

Kirp, David L. (2013). *Improbable Scholars: The Rebirth of a Great American School System and a Strategy for America's Schools*. Oxford University Press.

Kiva.org. (n.d.). About microfinance, www.kiva.org/about/microfinance.

Kohlberg, Lawrence, Charles Levine, and Alexandra Hewer. (1983). *Moral Stages: A Current Formulation and a Response to Critics*. Contributions to Human Development. Karger Publishing.

Kolbert, Elizabeth. (2012). Spoiled rotten: Why do kids rule the roost? *New Yorker*, July 2, 2012, www.newyorker.com/arts/critics/books/2012/07/02/120702crbo_books_kolbert.

Koltko-Rivera, Mark E. (2006). Rediscovering the later version of Maslow's hierarchy of needs: Self-transcendence and opportunities for theory, research, and unification. *Review of General Psychology* 10(4):203–317, http://psycnet.apa.org/doi/10.1037/1089-2680.10.4.302.

Koomey, Jonathan. (2011). Growth in data center electricity use 2005 to 2010. Analytics Press,www.mediafire.com/file/zzqna34282frr2f/koomeydatacenter electuse2011finalversion.pdf.

Kralev, Nicholas. (2009). India tells Clinton: No carbon cuts. *Washington Times*, July 20, 2009, www.washingtontimes.com/news/2009/jul/20/india-tells -clinton-no-carbon-cuts/.

Kranzberg, Melvin. (1986). Technology and history: "Kranzberg's laws." *Technology and Culture* 27(30):544–560, www.jstor.org/stable/3105385.

Kraus, Michael W., Stéphane Côté, and Dacher Keltner. (2010). Social class, contextualism, and empathic accuracy. *Psychological Science* 21:1716–1723, http://pss.sagepub.com/content/21/11/1716.

Krishna, Anirudh. (2010). *One Illness Away: Why People Become Poor and How They Escape Poverty*. Oxford University Press.

Kristof, Nicholas D., and Sheryl WuDunn. (2009). *Half the Sky: Turning Oppression into Opportunity for Women Worldwide*. Vintage.

Krueger, Alan B., and Mikael Lindahl. (2001). Education for growth: Why and for whom? *Journal of Economic Literature* 39(4):1101–1136, www.jstor.org /stable/2698521.

Krugman, Paul. (2010). Block those metaphors. *New York Times*, Dec. 12, 2010, www.nytimes.com/2010/12/13/opinion/13krugman.html.

Kumar, Divya. (2008). Study of split screen in shared-access scenarios: Optimizing value of PCs in resource-constrained classrooms in developing countries. Master's thesis, University of California, San Diego, https://escholarship.org/uc/item/7x8081jw.

Kuriyan, Renee, and Kentaro Toyama, eds. (2007). Review of research on rural PC kiosks. Microsoft Research, http://research.microsoft.com/en-us/um /india/projects/ruralkiosks/Kiosks%20Research.doc.

Lankarani, Nazanin. (2011). Transforming Africa through higher education. *New York Times*, Jan. 16, 2011, www.nytimes.com/2011/01/17/world/africa /17iht-educSide17.html.

Lareau, Annette. (2011). *Unequal Childhoods: Class, Race, and Family Life*, 2nd ed. University of California Press.

Larsen, Erling Røed. (2004). Escaping the resource curse and the Dutch disease? When and why Norway caught up with and forged ahead of its neighbors. Discussion Papers No. 377, May 2004. Statistics Norway, Research Department, www.ssb.no/publikasjoner/pdf/dp377.pdf.

Latour, Bruno. (1991). Technology is society made durable. Pp. 103–132 in J. Law, ed., *A Sociology of Monsters: Essays on Power, Technology and Domination*. Sociological Review Monograph No. 38, www.bruno-latour.fr/sites/default /files/46-TECHNOLOGY-DURABLE-GBpdf.pdf.

Layard, Richard. (2005). *Happiness: Lessons from a New Science*. Penguin.

Lee, Eric, and Benjamin Weinthal. (2011). Trade unions: The revolutionary social network at play in Egypt and Tunis. *The Guardian*, Feb. 10, 2011, www.theguardian.com/commentisfree/2011/feb/10/trade-unions-egypt-tunisia.

Lee, K., J. S. Brownstein, R. G. Mills, and I. S. Kohane. (2010). Does collocation inform the impact of collaboration? *PLOS ONE* 5(12):e14279, www.plosone.org/article/info%3Adoi%2F10.1371%2Fjournal.pone.0014279.

Lee, Meredith. (1998). Popular quotes: Commitment. Goethe Society of North America, www.goethesociety.org/pages/quotescom.html.

Leland, John. (2011). Out on the town, always online. *New York Times*, Nov. 19, 2011, www.nytimes.com/2011/11/20/nyregion/out-on-the-town-always-online.html.

Lemov, Doug. (2010). *Teach Like a Champion: 49 Techniques That Put Students on the Path to College (K-12)*. Jossey-Bass.

Lenhart, Amanda. (2012). Teens, smartphones & texting. Pew Research Center, March 19, 2012, www.pewinternet.org/files/old-media//Files/Reports/2012/PIP_Teens_Smartphones_and_Texting.pdf.

Levitt, Steven D., and Stephan J. Dubner. (2006). *Freakonomics: A Rogue Economist Explores the Hidden Side of Everything*, rev. ed. William Morrow.

Lewis, David (2005). Anthropology and development: The uneasy relationship. Pp. 472–486 in James G. Carrier, ed., *A Handbook of Economic Anthropology*. Edward Elgar, http://eprints.lse.ac.uk/253/1/Anthropology_and_development_a_brief_overview.pdf.

Lewis, Oscar. (1961). *The Children of Sanchez: Autobiography of a Mexican Family*. Vintage.

Lincoln, Abraham. (1858). First debate against Stephen Douglas. Ottawa, Illinois, Aug. 21, 1858.

Linden, Leigh L. (2008). Complement or substitute? The effect of technology on student achievement in India. Working Paper, www.leighlinden.com/Gyan_Shala_CAL_2008-06-03.pdf.

Linnell, Natalie, Richard Anderson, Guy Bordelon, Rikin Gandhi, Bruce Hemingway, S. B. Nadagouda, and Kentaro Toyama. (2011). Context-aware technology for improving interaction in video-based agricultural extension. *India HCI*, http://dl.acm.org/citation.cfm?id=2407799.

Lipset, Seymour Martin. (1959). Some social requisites of democracy: Economic development and political legitimacy. *American Political Science Review* 53: 69–105, http://dx.doi.org/10.2307/1951731.

———. (1960). *Political Man: The Social Bases of Politics*. Doubleday.

Lubow, Arthur. (2007). Conductor of the people. *New York Times*, Oct. 28, 2007, www.nytimes.com/2007/10/28/magazine/28dudamel-t.html.

Luntz, Frank I. (2009). *What Americans Really Want . . . Really: The Truth About Our Hopes, Dreams, and Fears*. Hyperion.

Lyubomirsky, Sonja. (2007). *The How of Happiness: A New Approach to Getting the Life You Want*. Penguin.

MacKenzie, Donald, and Judy Wajcman, eds. (1985). *The Social Shaping of Technology*. Open University Press.

Mahoney, Joseph L., Angel L. Harris, and Jacquelynne S. Eccles. (2006). Organized activity participation, Positive youth development, and the over-scheduling hypothesis. Society for Research in Child Development. *Social Policy Report* 20(16), www.srcd.org/sites/default/files/documents/spr_20-4.pdf.

Malmodin, J., Å. Moberg, D. Lundén, G. Finnveden, and N. Lövehagen. (2010). Greenhouse gas emissions and operational electricity use in the ICT and entertainment & media sectors. *Journal of Industrial Ecology*, http://onlinelibrary.wiley.com/doi/10.1111/j.1530-9290.2010.00278.x/abstract.

Mandela, Nelson. (2003). Lighting your way to a better future. Speech delivered at launch of Mindset Network, July 16, 2003, http://db.nelsonmandela.org/speeches/pub_view.asp?pg=item&ItemID=NMS909.

Mankiw, N. Gregory. (2004). *Principles of Economics*, 3rd ed. Thomson South-Western.

MarketLine. (2014). Global – Carbonated soft drinks. Aug. 19, 2014, http://store.marketline.com/Product/global_carbonated_soft_drinks?productid=MLIP1364-0008.

Maslow, Abraham. H. (1943). A theory of human motivation. *Psychological Review* 50(4):370–396, http://psycnet.apa.org/psycinfo/1943-03751-001.

———. (1954 [1987]). *Motivation and Personality*, 3rd ed., rev. by Robert Frager, James Fadiman, Cynthia McReynolds, and Ruth Cox. Addison-Wesley Educational Publishers.

———. (1961). Are our publications and conventions suitable for the personal sciences? *American Psychologist* 16:318–319, http://psycnet.apa.org/psycinfo/2005-14424-005.

———. (1965). *Eupsychian Management: A Journal*. Richard D. Irwin and Dorsey Press.

———. (1968). *Toward a Psychology of Being*, 2nd ed. D. Van Nostrand.

———. (1971). *The Farther Reaches of Human Nature*. Viking.

———. (1996). Critique of self-actualization theory. Pp. 26–32 in E. Hoffman, ed., *Future Visions: The Unpublished Papers of Abraham Maslow*. Sage.

Mbiti, Isaac, and David N. Weil. (2011). Mobile banking: The impact of M-PESA in Kenya. National Bureau of Economic Research Working Paper 17129, www.nber.org/papers/w17129.

McClelland, David C. (1961). *The Achieving Society*. Free Press.

McNeil, Donald G. (2010). A poor nation, with a health plan. *New York Times*, June 14, 2010, www.nytimes.com/2010/06/15/health/policy/15rwanda .html.

Medhi, Indrani, Meera Lakshmanan, Kentaro Toyama, and Edward Cutrell. (2013). Some evidence for the impact of limited education on hierarchical user interface navigation. Pp. 2813–2822 in *Proceedings of the SIGCHI Conference on Human Factors in Computing Systems (CHI '13)*. ACM, http://doi.acm.org/10.1145/2470654.2481390.

Medhi, Indrani, Archana Prasad, and Kentaro Toyama. (2007). Optimal audio-visual representations for illiterate users of computers. Pp. 873–882 in *Proceedings of the 16th International Conference on World Wide Web (WWW '07)*. ACM, http://doi.acm.org/10.1145/1242572.1242690.

Merritt, Anna C., Daniel A. Effron, and Benoît Monin. (2010). Moral self-licensing: When being good frees us to be bad. *Social and Personality Psychology Compass* 4/5:344–357, http://onlinelibrary.wiley.com/doi /10.1111/j.1751-9004.2010.00263.x/abstract.

Milliot, Jim. (2014). Book sales rose 1% in 2013. *Publishers Weekly*, April 1, 2014, www.publishersweekly.com/pw/by-topic/industry-news/financial -reporting/article/61667-book-sales-rose-1-in-2013.html.

Mischel, Walter, and Yuichi Shoda. (1995). A cognitive-affective system theory of personality: Reconceptualizing situations, dispositions, dynamics, and invariance in personality structure. *Psychological Review* 102(2):246–268, http://psycnet.apa.org/journals/rev/102/2/246/.

Mitra, Sugata, and Payal Arora. (2010). Afterthoughts. *British Journal of Educational Technology* 41(5):703–705, http://onlinelibrary.wiley.com /doi/10.1111/j.1467-8535.2010.01079.x/abstract.

Mitra, Sugata, and Ritu Dangwal. (2010). Limits to self-organising systems of learning – the Kalikuppam experiment. *British Journal of Educational Technology* 41(5):672–688, www.hole-in-the-wall.com/docs/Paper13.pdf.

MixMarket. (2014). Microfinance institutions, www.mixmarket.org/mfi/.

Mnookin, Seth. (2011). *The Panic Virus: A True Story of Medicine, Science, and Fear*. Simon and Schuster.

Morawczynski, Olga. (2011). Examining the adoption, usage and outcomes of mobile money services: The case of M-PESA in Kenya. PhD Thesis, University of Edinburgh, https://www.era.lib.ed.ac.uk/bitstream/1842 /5558/2/Morawczynski2011.pdf.

Morawczynski, Olga, and Mark Pickens. (2009). Poor people using mobile financial services: Observations on customer usage and impact from M-PESA. *CGAP Brief*, Aug. 2009, https://www.cgap.org/sites/default/files/CGAP-Brief-Poor-People-Using-Mobile-Financial-Services-Observations-on-Customer-Usage-and-Impact-from-M-PESA-Aug-2009.pdf.

Morozov, Evgeny. (2011). *The Net Delusion: The Dark Side of Internet Freedom*. PublicAffairs.

———. (2013). *To Save Everything Click Here: The Folly of Technological Solutionism*. PublicAffairs.

Mueller, Claudia M., and Carol S. Dweck. (1998). Praise for intelligence can undermine children's motivation and performance. *Journal of Personality and Social Psychology* 75:33–52, http://psycnet.apa.org/doi/10.1037/0022-3514.75.1.33.

Mukul, Akshaya. (2006). HRD rubbishes MIT's laptop scheme for kids. *Times of India*, July 3, 2006, http://articles.timesofindia.indiatimes.com/2006-07-03/india/27814789_1_hrd-ministry-million-laptops-laptop-scheme.

Mumford, Lewis. (1966). *The Myth of the Machine: Technics and Human Development*. Harcourt, Brace and World.

Murphy, Tom. (2014a). Do TOMS shoes harm local shoe sellers? *Humanosphere*, Sept. 16, 2014, www.humanosphere.org/social-business/2014/09/toms-shoes-harm-local-shoe-sellers/.

———. (2014b). Is this the nail in the One Laptop Per Child coffin? *Humanosphere*, Sept. 30, 2014, www.humanosphere.org/basics/2014/09/nail-one-laptop-per-child-coffin/.

Narayan, Deepa, Robert Chambers, Meera Kaul Shah, and Patti Petesch. (2000). *Voices of the Poor*, vols. 1–3. Published for the World Bank. Oxford University Press.

National Center for Charitable Statistics. (2010). Number of nonprofit organizations in the United States, 1999–2009, http://nccsdataweb.urban.org/PubApps/profile1.php?state=US.

National Coalition for Child Protection Reform. (2003). Poverty is the leading cause of child abuse. In Louise I. Gerdes, ed., *Child Abuse: Opposing Viewpoints*. Greenhaven.

Negroponte, Nicholas. (2008). Taking OLPC to Colombia. TED in the Field, Dec. 2008, www.ted.com/talks/nicholas_negroponte_takes_olpc_to_colombia.html.

Neher, Andrew. (1991). Maslow's theory of motivation: A critique. *Journal of Humanistic Psychology* 31(3):89–112, http://jhp.sagepub.com/cgi/content/abstract/31/3/89.

New York Times. (2012). Transcript of the First Presidential Debate. Oct. 3, 2012, www.nytimes.com/2012/10/03/us/politics/transcript-of-the-first-presidential-debate-in-denver.html.

Nisbet, Robert. (1980). *History of the Idea of Progress*. Basic Books.

Nussbaum, Martha C. (2011). *Creating Capabilities: The Human Development Approach*. Belknap Press of Harvard University Press.

Obama, Barack. (2013). Inaugural Address by President Barack Obama. Whitehouse.gov, Jan. 21, 2013, www.whitehouse.gov/the-press-office/2013/01/21/inaugural-address-president-barack-obama.

O'Connor, Clare. (2014). Bain deal makes TOMS Shoes founder Blake Mycoskie a $300 million man. *Forbes*, Aug. 20, 2014, www.forbes.com/sites/clareoconnor/2014/08/20/bain-deal-makes-toms-shoes-founder-blake-mycoskie-a-300-million-man/.

Olivarez-Giles, Nathan. (2011). Libya's Internet reportedly down as violence against anti-government protesters continues. *Los Angeles Times*, Feb. 18, 2011, http://latimesblogs.latimes.com/technology/2011/02/libya-has-shut-down-the-internet-in-light-of-protests-reports-say.html.

Olson, Gary M., and Judith S. Olson. (2000). Distance matters. *Human-Computer Interaction* 15:139–178, http://dl.acm.org/citation.cfm?id=1463019.

One Laptop Per Child. (n.d.). Mission, http://laptop.org/en/vision/mission/.

Oppenheimer, Todd. (2003). *The Flickering Mind: Saving Education from the False Promise of Technology*. Random House.

Opportunity International. (n.d.). Microfinance – a working solution to global poverty (an introduction), http://opportunity.org/about-us/opportunity-international-video/microfinance-a-working-solution-to-global-poverty-(an-introduction).

Organisation for Economic Co-operation and Development (OECD). (2010a). PISA 2009 results: Executive summary. OECD Publishing, www.oecd.org/pisa/pisaproducts/46619703.pdf.

———. (200b). PISA 2009 results: What makes a school successful? vol. 4. OECD Publishing, www.oecd.org/pisa/pisaproducts/48852721.pdf.

———. (2011). Lessons from PISA for the United States: Strong performers and successful reformers in education. OECD Publishing, http://dx.doi.org/10.1787/9789264096660-en.

———. (2013a). Life expectancy at birth. In *Health at a glance 2013: OECD indicators*. OECD Publishing, http://dx.doi.org/10.1787/health_glance-2013-5-en.

———. (2013b). PISA 2012 results: Excellence through equity: Giving every student the chance to succeed, vol. 2. OECD Publishing, http://dx.doi.org/10.1787/9789264201132-en.

———. (2014a). Aid to education. OECD Publishing, Oct. 2014, www.oecd.org/dac/stats/documentupload/Aid%20to%20Education%20data%20to%202011-12.pdf.

———. (2014b). PISA 2012 results: What students know and can do: Student performance in mathematics, reading and science, vol. 1, rev. ed., OECD Publishing, Feb. 2014, http://dx.doi.org/10.1787/9789264208780-en.

Orwell, George. (1936). Review of Penguin Books. *New English Weekly*, March 5, 1936.

O'Toole, Garson. (2010). Not everything that counts can be counted. Quote Investigator blog, May 26, 2010, http://quoteinvestigator.com/2010/05/26/everything-counts-einstein/.

———. (2012). It is not enough to succeed; one's best friend must fail. Quote Investigator blog, Aug. 6, 2012, http://quoteinvestigator.com/2012/08/06/succeed-fail/.

———. (2013). The mind is not a vessel that needs filling, but wood that needs igniting. Quote Investigator blog, March 28, 2013, http://quoteinvestigator.com/2013/03/28/mind-fire/.

Oxford English Dictionary, 2nd ed. (2013). OED Online, Dec. 2013, www.oed.com/.

Packer, George. (2013). Change the world. *New Yorker*, May 27, 2013, www.newyorker.com/magazine/2013/05/27/change-the-world.

———. (2014). Cheap words. *New Yorker*, Feb. 17, 2014, http://www.newyorker.com/magazine/2014/02/17/cheap-words.

Page, Larry. (2014). Where's Google going next? Interview with Charlie Rose. TED 2014, www.ted.com/talks/larry_page_where_s_google_going_next.

Paiva, F. J. X. (1977). A conception of social development. *Social Science Review* 51(2):327–336, www.jstor.org/stable/30015486.

Pal, Joyojeet. (2005). Computer aided learning survey. Microsoft Research, http://research.microsoft.com/en-us/um/india/projects/computeraidedlearningsurvey/index.htm.

Pal, Joyojeet, Meera Lakshmanan, and Kentaro Toyama. 2009. "My child will be respected": Parental perspectives on computers and education in rural India. *Information Systems Frontiers* 11(2):129–144, http://dx.doi.org/10.1007/s10796-009-9172-1.

Pal, Joyojeet, Udai Singh Pawar, Eric A. Brewer, and Kentaro Toyama. (2006). The case for multi-user design for computer aided learning in developing regions. Pp. 781–789 in *Proceedings of the 15th international Conference on*

World Wide Web. Edinburgh, May 23–26, 2006. ACM, http://doi.acm.org/10.1145/1135777.1135896.

Panjwani, Saurabh, Aakar Gupta, Navkar Samdaria, Edward Cutrell, and Kentaro Toyama. (2010). Collage: A presentation tool for school teachers. In *Proceedings of the International Conference on Information and Communication Technologies and Development (ICTD 2010)*. London, http://dl.acm.org/citation.cfm?id=2369248.

Parfit, Derek. (1984). *Reasons and Persons*. Oxford University Press.

Park, Madison. (2014). Top 20 most polluted cities in the world. CNN, May 8, 2014, www.cnn.com/2014/05/08/world/asia/india-pollution-who/.

Parsons, Christ, Julie Makienen, and Neela Banerjee. (2014). Obama, Chinese president agree to landmark climate deal. *Los Angeles Times*, Nov. 12, 2014, www.latimes.com/world/asia/la-fg-obama-xi-climate-change-20141111-story.html.

Paruthi, Gaurav, William Thies. (2011). Utilizing DVD players as low-cost offline Internet browsers. In *ACM Conference on Human Factors in Computing Systems (CHI 2011)*, http://research.microsoft.com/en-us/um/people/thies/chi11-thies.pdf.

Patrinos, Harry Anthony. (2008). Returns to education: The gender perspective. Pp. 53–66 in Mercy Temblon and Lucia Fort, eds., *Girls' Education in the 21st Century*. World Bank.

Paulhus, Delroy L., Paul Wehr, Peter D. Harms, and David I. Strasser. (2002). Use of exemplar surveys to reveal implicit types of intelligence. Leadership Institute Faculty Publications Paper 15, http://digitalcommons.unl.edu/leadershipfacpub/15.

Paul Revere Heritage Project. (n.d.). One if by land, two if by sea, www.paul-revere-heritage.com/one-if-by-land-two-if-by-sea.html.

Paulson, Tom. (2011). Geek heretic: Technology cannot end poverty. *Humanosphere*, May 6, 2011, www.humanosphere.org/science/2011/05/geek-heretic-technology-cannot-end-poverty/.

Pawar, Udai Singh, Joyojeet Pal, Rahul Gupta, and Kentaro Toyama. (2007). Multiple mice for retention tasks in disadvantaged schools. Pp. 1581–1590 in *Proceedings of the SIGCHI Conference on Human Factors in Computing Systems (CHI '07)*. ACM, http://doi.acm.org/10.1145/1240624.1240864.

Pawson, Ray. (2004). Mentoring relationships: An explanatory review. ESRC UK Centre for Evidence Based Policy and Practice Working Paper 21, www.kcl.ac.uk/sspp/departments/politicaleconomy/research/cep/pubs/papers/assets/wp21.pdf.

Perry B. E. (1990). *Babrius and Phaedrus*. Loeb Classical Library. Harvard University Press.

Perry, Suzanne. (2013). The stubborn 2% giving rate. *Chronicle of Philanthropy*, June 17, 2013, http://philanthropy.com/article/The-Stubborn-2-Giving-Rate/139811/.

Peterson, Christopher, and Martin E. P. Seligman. (2004). *Character Strengths and Virtues: A Handbook and Classification*. Oxford University Press.

Piaget, Jean, and Baerbel Inhelder. (1958). *The Growth of Logical Thinking from Childhood to Adolescence*. Basic Books.

Piff, Paul K., Michael W. Kraus, Stéphane Côté, Bonnie Hayden Cheng, and Dacher Keltner. (2010). Having less, giving more: The influence of social class on prosocial behavior. *Journal of Personality and Social Psychology* 99(5):771–784, http://psycnet.apa.org/journals/psp/99/5/771/.

Piff, Paul K., Daniel M. Stancato, Stéphane Côté, Rodolfo Mendoza-Denton, Dacher Keltner. (2012). Higher social class predicts increased unethical behavior. In *Proceedings of the National Academy of Sciences*, www.pnas.org/content/early/2012/02/21/1118373109.

Piketty, Thomas. (2014). *Capital in the Twenty-First Century*. Arthur Goldhammer, trans. Belknap Press of Harvard University Press.

Piketty, Thomas, Emmanuel Saez. (2003). Income inequality in the United States, 1913–1998. *Quarterly Journal of Economics* 143(1):1–39, http://qje.oxfordjournals.org/content/118/1/1.

Pinker, Steven. (2011). *The Better Angels of Our Nature: Why Violence Has Declined*. Viking.

Plato. (1956). *Great Dialogues of Plato*. W. H. D. Rouse, trans. Mentor Books, New American Library.

Plumer, Brad. (2012). What cookstoves tell us about the limits of technology. *Washington Post*, May 8, 2012, www.washingtonpost.com/blogs/wonkblog/post/what-cook-stoves-tell-us-about-the-limits-of-technology/2012/05/08/gIQApp8YAU_blog.html.

Plutarch. (1992). *Essays*. Robin Waterfield, trans. Penguin Classics.

Polgreen, Lydia, and Vikas Bajaj. (2010). India microcredit faces collapse from defaults. *New York Times*, Nov. 17, 2010, www.nytimes.com/2010/11/18/world/asia/18micro.html.

Population Research Institute. (1998). Fact sheet on sterilization campaigns around the world. Congressional Briefing, Feb. 23, 1998, www.pop.org/content/fact-sheet-on-sterilization-campaigns-around-the-world-872.

Porter, Eduardo. (2013). To address gender gap, is it enough to lean in? *New York Times*, Sept. 24, 2013, www.nytimes.com/2013/09/25/business/economy/for-american-women-is-it-enough-to-lean-in.html.

Postman, Neil. (1985 [2005]). *Amusing Ourselves to Death: Public Discourse in the Age of Show Business*, 20th Anniversary Edition. Penguin.

Pradan. (2014). PRADAN annual report 2013–2014, www.pradan.net/index.php?option=com_content&task=view&id=109&Itemid=88.

———. (n.d.). Mission, www.pradan.net/index.php?option=com_content&task=view&id=18&Itemid=4.

Prahalad, C. K. (2004). *The Fortune at the Bottom of the Pyramid: Eradicating Poverty Through Profits*. Wharton School Publishing.

Prensky, Marc. (2011). Digital natives, digital immigrants. In Marc Bauerlein, ed., *The Digital Divide: Arguments for and Against Facebook, Google, Texting, and the Age of Social Networking*. Tarcher/Penguin.

Pritchett, Lant. (1996). Where Has All the Education Gone? Policy Working Research Paper 1581. World Bank, http://unpan1.un.org/intradoc/groups/public/documents/UNPAN/UNPAN002390.pdf.

Przybylski, Andrew K., Kou Murayama, Cody R. DeHaan, and Valerie Gladwell. (2013). Motivational, emotional, and behavioral correlates of fear of missing out. *Computers in Human Behavior* 29(4):1814–1848, www.sciencedirect.com/science/article/pii/S0747563213000800.

Psacharpoulos, George, and Harry Anthony Patrinos. (2004). Returns to investment in education: A further update. *Education Economics* 12(2):111–134, http://elibrary.worldbank.org/doi/pdf/10.1596/1813-9450-2881.

Raina, Pamposh, and Heather Timmons. (2011). Meet Aakash, India's $35 'laptop.' *New York Times*, Oct. 5, 2011, http://india.blogs.nytimes.com/2011/10/05/meet-aakash-indias-35-laptop.

Ramkumar, Vivek. (2008). Our money, our responsibility: A citizens' guide to monitoring government expenditures. International Budget Project, http://internationalbudget.org/wp-content/uploads/Our-Money-Our-Responsibility-A-Citizens-Guide-to-Monitoring-Government-Expenditures-English.pdf.

Rangaswamy, Nimmi. (2009). The non-formal business of cyber cafés: A case-study from India. *Journal of Information, Communication and Ethics in Society* 7(2/3):136–145, www.emeraldinsight.com/doi/abs/10.1108/14779960910955855.

Rao, Leena. (2011). More evidence that Facebook is nearing 600 million users. *Tech Crunch*, Jan. 13, 2011, http://techcrunch.com/2011/01/13/facebook-nearing-600-million-users/.

Ratan, Aishwarya Lakshmi, Sambit Satpathy, Lilian Zia, Kentaro Toyama, Sean Blagsvedt, Udai Singh Pawar, and Thanuja Subramaniam. (2009). Kelsa+: Digital literacy for low-income office workers. In *Proceedings on the International Conference on Information and Communication Technologies and Development*. IEEE Press, http://dx.doi.org/10.1109/ICTD.2009.5426713.

Ravallion, Martin. (2011). Guest post by Martin Ravallion: Are we really assessing development impact? Development Impact blog, World Bank, May 25, 2011, http://blogs.worldbank.org/impactevaluations/guest-post-by-martin-ravallion-are-we-really-assessing-development-impact.

Ravitch, Diane. (2011). *The Death and Life of the Great American School System: How Testing and Choice Are Undermining Education*, rev ed. Basic Books.

Ray, Daniel P., and Yasmin Ghahremani. (2014). Credit card statistics, industry facts, debt statistics. CreditCards.com, www.creditcards.com/credit-card-news/credit-card-industry-facts-personal-debt-statistics-1276.php.

Reinhardt, Uwe. (2012). Divide et impera: Protecting the growth of health care incomes (costs). *Health Economics* 21:41–54, http://onlinelibrary.wiley.com/doi/10.1002/hec.1813/abstract.

Resources Aimed at the Prevention of Child Abuse and Neglect. (1997). Child abuse and the impact of poverty. Agenda Feminist Media. *Agenda: Empowering Women for Gender Equity* 33:43–48, www.jstor.org/stable/4066132.

Revkin, Andrew. (2008). Norway hikes aid despite economy. Dot Earth blog, *New York Times*, Oct. 8, 2008, http://dotearth.blogs.nytimes.com/2008/10/08/norway-hikes-aid-despite-economy/.

Rhodes, Jean E. (2008). Improving youth mentoring interventions through research-based practice. *American Journal of Community Psychology* 41 (1–2):35–42, http://link.springer.com/article/10.1007%2Fs10464-007-9153-9.

Roodman, David. (2012). *Due Diligence: An Impertinent Inquiry into Microfinance*. Center for Global Development.

Rosenberg, Richard. (2007). CGAP reflections on the Compartamos initial public offering: A case study on microfinance interest rates and profits. CGAP Focus Note No. 42, June 2007, www.cgap.org/sites/default/files/CGAP-Focus-Note-CGAP-Reflections-on-the-Compartamos-Initial-Public-Offering-A-Case-Study-on-Microfinance-Interest-Rates-and-Profits-Jun-2007.pdf.

Rosenfeld, M. J., and Reuben J. Thomas. (2012). Searching for a mate: The rise of the Internet as a social intermediary. *American Sociological Review* 77(4):523–547, http://asr.sagepub.com/content/77/4/523.

Rossi, Peter H. (1987). The iron law of evaluation and other metallic rules. *Research in Social Programs and Public Policy* 4:3–20.

Rostow, W. W. (1960). *The Stages of Economic Growth: A Non-Communist Manifesto*. Cambridge University Press.

Rowan, David. (2010). Kinect for Xbox 360: The inside story of Microsoft's secret "Project Natal." *Wired UK*, Oct. 29, 2010, www.wired.co.uk /magazine/archive/2010/11/features/the-game-changer.

Rowan, John. (1998). Maslow amended. *Journal of Humanistic Psychology* 38(1):81–92, http://jhp.sagepub.com/content/38/1/81.short.

Rowe, Jonathan. (2008). Our phony economy. *Harper's*, June 2008, http://harpers.org/archive/2008/06/our-phony-economy/.

Rubin, Gretchen. (2009). *The Happiness Project: Or, Why I Spent a Year Trying to Sing in the Morning, Clean My Closets, Fight Right, Read Aristotle, and Generally Have More Fun*. Harper.

Rumpf, Matthias, Sarah Clune, and Jason Kane. (2011). Why does health care cost so much in the United States? *NPR Newshour*, Nov., 25, 2011, www.pbs.org /newshour/rundown/2011/11/why-does-healthcare-cost-so-much.html.

Rupp, Lindsey, and Devin Banerjee. (2014). Toms sells 50% stake to Bain Capital to fund sales growth. *Bloomberg*, Aug. 20, 2014, www.bloomberg. com/news/2014-08-20/toms-sells-50-stake-to-bain-capital.html.

Sachs, Jeffrey. (2005). *The End of Poverty: How We Can Make It Happen in Our Lifetime*. Penguin.

———. (2008). The digital war on poverty. *Project Syndicate*, Aug. 20, 2008, www.project-syndicate.org/commentary/the-digital-war-on-poverty.

Sachs, Jeffrey D., and Andrew M. Warner. (1999). The big rush, natural resource booms and growth. *Journal of Development Economics* 59(1):43–76, www.sciencedirect.com/science/article/B6VBV-3WMK4TP-3/2/7e2c00 30bf45b0f9a0d8cd5b8cbec71e.

Saez, Emmanuel. (2013). Striking it richer: The evolution of top incomes in the United States. UC Berkeley, http://eml.berkeley.edu/~saez/saez-UStopincomes-2012.pdf.

Sahni, Urvashi, Rahul Gupta, Glynda Hull, Paul Javid, Tanuja Setia, Kentaro Toyama, and Randy Wang. (2008). Using digital video in rural Indian schools: A study of teacher development and student achievement. In *Annual Meeting of the American Educational Research Association*, March 2008, http://dsh.cs.washington.edu:8000/distance/08aAERA.pdf.

Salamon, Lester M., S. Wojciech Sokolowski, Megan Haddock, and Helen S. Tice. (2013). The state of global civil society and volunteering: Latest findings from the implementation of the UN Nonprofit Handbook. Johns Hopkins Center for Civil Society Studies, Working Paper #49, March 2013, http://ccss.jhu.edu/wp-content/uploads/downloads/2013/04/JHU_Global -Civil-Society-Volunteering_FINAL_3.2013.pdf.

Sandberg, Sheryl. (2014). *Lean In for Graduates*. Knopf.

Sandel, Michael J. (2012). *What Money Can't Buy: The Moral Limits of Markets*. Farrar, Straus and Giroux.

Sanders, Sam H. (2013). Students find ways to hack school-issued iPads within a week. *NPR All Tech Considered*, Sept. 27, 2013, www.npr.org/blogs/all techconsidered/2013/09/27/226654921/ students-find-ways-to-hack-school-issued-ipads-within-a-week.

Sanderson, Susan, and Mustafa Uzumeri. (1995). Managing product families: The case of the Sony Walkman. *Research Policy* 24:761–782, www.sciencedirect.com/science/article/pii/004873339400797B.

Santiago, Ana, E. Severin, J. Cristia, P. Ibarrarán, J. Thompson, and S. Cueto. (2010). Evaluacíon experimental del programa "Una Laptop por Niño" en Perú. Banco Interamericano de Desarrollo, www.iadb.org/document. cfm?id=35370099.

Sartre, Jean-Paul. (1957 [1983]). *Existentialism and Human Emotions*. Hazel E. Barnes, trans. Citadel.

Sawyer, Kathy. (1999). Armstrong's code. *Washington Post Magazine*, July 11, 1999, www.washingtonpost.com/wp-srv/national/longterm/space/armstrong full.htm.

Saxenian, AnnaLee. (2006). *The New Argonauts: Regional Advantage in a Global Economy*. Harvard University Press.

Saxenian, AnnaLee, Geoffrey Nunberg, Eric Brewer, Megan Smith, Kentaro Toyama, and Wayan Vota. (2011). Digital divide or digital bridge: Can information technology alleviate poverty? Panel at School of Information, University of California, Berkeley, April 6, 2011, www.ischool.berkeley.edu /newsandevents/events/technologyandpoverty2011.

Scheer, Roddy, and Doug Moss. (2012). Use it and lose it: The outsize effect of U.S. consumption on the environment. Earth Talk, *Scientific American*, Sept. 14, 2012, www.scientificamerican.com/article/american-consumption -habits/.

Schmidt, Eric, and Jared Cohen. (2013). *The New Digital Age: Reshaping the Future of People, Nations and Business*. Alfred A. Knopf.

Schramm, Wilbur. (1964). *Mass Media and National Development: The Role of Information in the Developing Countries*. Stanford University Press.

Schwartz, Barry, and Kenneth Sharpe. (2010). *Practical Wisdom: The Right Way to Do the Right Thing*. Riverhead Books.

Schwarz, Norbert, and Fritz Strack. (1999). Reports of subjective well-being: Judgmental processes and their methodological implications. Pp. 61–84 in *Well-Being: The Foundations of Hedonic Psychology*. Russell Sage Foundation.

Scott, Christopher D. (2000). Mixed fortunes: A study of poverty mobility among small farm households in Chile, 1968–86. *Journal of Development*

Studies 36(6):155–180, www.tandfonline.com/doi/abs/10.1080/00220 380008422658.

Seligman, Martin E. P. (1975). *Helplessness: On Depression, Development, and Death*. W. H. Freeman.

———. (2002). *Authentic Happiness: Using the New Positive Psychology to Realize Your Potential for Lasting Fulfillment*. Free Press.

———. (2006). *Learned Optimism: How to Change Your Mind and Your Life*. Vintage.

Seligman, Martin E. P., and S. F. Maier. (1967). Failure to escape traumatic shock. *Journal of Experimental Psychology* 74:1–9, http://psycnet.apa.org /psycinfo/1967-08624-001.

Sen, Amartya. (2000). *Development as Freedom*. Oxford University Press.

Service, Elman R. (1962 [1968]). *Primitive Social Organization: An Evolutionary Perspective*. Random House.

Sey, Araba, and Michelle Fellows. (2009). Literature review on the impact of public access to information and communication technologies. TASCHA Working Paper No. 6. Technology & Social Change Group, University of Washington, Seattle, http://library.globalimpactstudy.org/sites/default/files /docs/CIS-WorkingPaperNo6.pdf.

Sheldon, Kennon M., and Andrew J. Elliot. (1998). Not all personal goals are personal: Comparing autonomous and controlled reasons as predictors of effort and attainment. *Personality and Social Psychology Bulletin* 24: 546–557, http://psp.sagepub.com/content/24/5/546.short.

———. (1999). Goal striving, need satisfaction, and longitudinal well-being: The self-concordance model. *Journal of Personality and Social Psychology* 76(3):482–497, http://psycnet.apa.org/journals/psp/76/3/482/.

Sheldon, Kennon M., and L. Houser-Marko. (2001). Self-concordance, goal attainment, and the pursuit of happiness: Can there be an upward spiral? *Journal of Personality and Social Psychology* 80: 152–165, http://psycnet.apa .org/journals/psp/80/1/152/.

Sheldon, Kennon M., and T. Kasser. (1998). Pursuing personal goals: Skills enable progress but not all progress is beneficial. *Personality and Social Psychology Bulletin* 24:1319–1331, http://psp.sagepub.com/content/24/12 /1319.short.

Shirky, Clay. (2010). *Cognitive Surplus: How Technology Makes Consumers into Collaborators*. Penguin.

———. (2011). The political power of social media. *Foreign Affairs*, Jan./Feb. 2011, www.foreignaffairs.com/articles/67038/clay-shirky/the-political-power -of-social-media.

———. (2014). Why I just asked my students to put their laptops away. Medium.com, https://medium.com/@cshirky/why-i-just-asked-my-students-to-put-their-laptops-away-7f5f7c50f368.

Shoda, Yuichi, Walter Mischel, and Philip K. Peake. (1990). Predicting adolescent cognitive and self-regulatory competencies from preschool delay of gratification: Identifying diagnostic conditions. *Developmental Psychology* 26(6):978–986, http://psycnet.apa.org/psycinfo/1991-06927-001.

Sinclair, Upton. (1934 [1994]). *I, Candidate for Governor: And How I Got Licked.* University of California Press.

Singer, Peter. (2009). *The Life You Can Save: Acting Now to End World Poverty.* Picador.

———. (2011). *The Expanding Circle: Ethics, Evolution, and Moral Progress.* Princeton University Press.

60 Minutes. (2011). Extra: Revolution 2.0. CBS News, Feb. 14, 2011, www.cbsnews.com/video/watch/?id=7349173n.

Small, Mario Luis, David J. Harding, and Michèle Lamont. (2010). Reconsidering culture and poverty. *Annals of the American Academy of Political and Social Science* 629:6–27, http://ann.sagepub.com/content/629/1/6.extract.

Smith, Hedrick. (2013). *Who Stole the American Dream?* Random House.

Smyth, Thomas N., Satish Kumar, Indrani Medhi, and Kentaro Toyama. (2010). Where there's a will there's a way: Mobile media sharing in urban India. In *Proceedings of the 28th International Conference Extended Abstracts on Human Factors in Computing Systems (CHI '10)*. ACM, http://dl.acm.org/citation.cfm?id=1753436.

Sobal, J., and A. J. Stunkard. (1989). Socioeconomic status and obesity: A review of the literature. *Psychological Bulletin* 105:260–275, http://content.apa.org/journals/bul/105/2/260.

Spiceworks. (2014). State of IT, http://itreports.spiceworks.com/reports/spiceworks_state-of-it-report-2014_print.pdf.

Star Trek: First Contact. (1996). Distributed by Paramount Pictures.

Stecklow, Steve. (2005). The $100 laptop moves closer to reality. *Wall Street Journal*, Nov. 14, 2005, http://online.wsj.com/news/articles/SB113193305149696140.

Steele, Claude M., and Joshua Aronson. (1995). Stereotype threat and the intellectual test performance of African Americans. *Journal of Personality and Social Psychology* 69(5):797–811, http://psycnet.apa.org/journals/psp/69/5/797/.

Stiglitz, Joseph E., Amartya Sen, and Jean-Paul Fitoussi, eds. (2009). Report by the Commission on the Measurement of Economic Performance and Social Progress, www.stiglitz-sen-fitoussi.fr/documents/rapport_anglais.pdf.

Stone, Linda. (2008). Continuous partial attention – not the same as multi-tasking. *Bloomberg Businessweek*, July 24, 2008, www.businessweek.com/business_at_work/time_management/archives/2008/07/continuous_part.html.

Strawson, Galen. (2010). *Freedom and Belief.* Oxford University Press.

Streitfeld, David. (2014). Dispute between Amazon and Hachette takes an Orwellian turn. *New York Times*, Aug. 9, 2014, http://bits.blogs.nytimes.com/2014/08/09/orwell-is-amazons-latest-target-in-battle-against-hachette/.

Surana, Sonesh, Rabin Patra, Sergiu Nedevschi, and Eric Brewer. (2008). Deploying a rural wireless telemedicine system: Experiences in sustainability. *IEEE Computer* 41(6):48–56, http://ieeexplore.ieee.org/xpl/articleDetails.jsp?arnumber=4548173.

Swaminathan, M. S. (2005). Mission 2007: Every village a knowledge centre. *The Hindu*, Nov. 25, 2005, www.thehindu.com/2005/11/25/stories/2005112504941000.htm.

Tabellini, Guido. (2008). Presidential address: Institutions and culture. *Journal of the European Economic Association* 6(2–3):255–294, www.jstor.org/stable/40282643.

Takahashi, M., and W. F. Overton. (2005). Cultural foundations of wisdom: An integrated developmental approach. In R. Sternberg and J. Jordan, eds., *A Handbook of Wisdom: Psychological Perspectives*. Cambridge University Press.

Tangney, June P., Roy F. Baumeister, and Angie Luzio Boone. (2004). High self-control predicts good adjustment, less pathology, better grades, and interpersonal success. *Journal of Personality* 72(2):271–324, http://onlinelibrary.wiley.com/doi/10.1111/j.0022-3506.2004.00263.x/abstract.

Taylor, Chris. (2011). Why not call it a Facebook Revolution? CNN, Feb. 24, 2011, www.cnn.com/2011/TECH/social.media/02/24/facebook.revolution/.

Taylor, William C. (1999). Inspired by work. *Fast Company*, Nov. 1999, www.fastcompany.com/38466/inspired-work.

Thaler, Richard H., and Cass R. Sunstein. (2008). *Nudge: Improving Decisions About Health, Wealth, and Happiness*. Yale University Press.

Thiel, Peter. (2012). Technology and regulation 3-3-12. The Federalist Society, https://www.youtube.com/watch?v=DDSO36mzBss.

Thompson, John B. (2010). *Merchants of Culture: The Publishing Business in the Twenty-First Century*. Polity.

Tichenor, P. J., G. A. Donohue, and C. N. Olien. (1970). Mass media flow and differential growth in knowledge. *Public Opinion Quarterly* 34(2):159–170, http://poq.oxfordjournals.org/content/34/2/159.abstract.

Toms. (n.d.). Toms Official Store, Toms.com, www.toms.com.

Toyama, Kentaro. (2010). Can technology end poverty? *Boston Review* 36(5):12–18, 28–29, www.bostonreview.net/forum/can-technology-end-poverty.

———. (2011). There are no technology shortcuts to good education. *Educational Technology Debate*, Jan. 2011, https://edutechdebate.org/ict-in-schools/there-are-no-technology-shortcuts-to-good-education/.

———. (2012). Q&A: The culture of medicine runs on people power, not tech. *The Atlantic*, June 27, 2012, www.theatlantic.com/health/archive/2012/06/q-a-the-culture-of-medicine-runs-on-people-power-not-tech/259000/.

———. (2013a). Our future might be bright: The tentative, rosy predictions of Google's Eric Schmidt. *The Atlantic*, May 2, 2013, www.theatlantic.com/technology/archive/2013/05/our-future-might-be-bright-the-tentative-rosy-predictions-of-googles-eric-schmidt/275360/.

———. (2013b). How Internet censorship actually works in China. *The Atlantic*, Oct. 2, 2013, www.theatlantic.com/china/archive/2013/10/how-internet-censorship-actually-works-in-china/280188/.

Toyama, Kentaro, and Andrew Blake. (2001). Probabilistic tracking in a metric space. In *Proceedings of the Eighth International Conference on Computer Vision* 2:50–57, http://dx.doi.org/10.1109/ICCV.2001.937599.

Tripathi, Salil. (2006). Microcredit won't make poverty history. *The Guardian*, Oct. 17, 2006, www.theguardian.com/business/2006/oct/17/business comment.internationalaidanddevelopment.

Tsotsis, Alexia. (2011). To celebrate the #Jan25 Revolution, Egyptian names his firstborn "Facebook." *Tech Crunch*, Feb. 19, 2011, http://techcrunch.co/2011/02/19/facebook-egypt-newborn/.

Tunstall, Tricia. (2012). *Changing Lives: Gustavo Dudamel, El Sistema, and the Transformative Power of Music*. Norton.

Turkle, Sherry. (2011). *Alone Together: Why We Expect More from Technology and Less from Each Other*. Basic Books.

Twenge, Jean M. (2006). *Generation Me: Why Today's Young Americans Are More Confident, Assertive, Entitled – and More Miserable Than Ever Before*. Free Press.

Uchitelle, Louis. (2008). Economists look to expand GDP to count "quality of life." *New York Times*, Sept. 1, 2008, www.nytimes.com/2008/09/01/business/worldbusiness/01iht-gdp.4.15791492.html.

UNESCO. (2012). Education for all global monitoring report 2012: Youth and skills – putting education to work, http://unesdoc.unesco.org/images/0021/002180/218003e.pdf.

United Nations. (2005). Annan unveils rugged $100 laptop for world's children at Tunis Summit. United Nations News Centre, Nov. 17, 2005, www.un.org/apps/news/story.asp?NewsID=16601.

———. (2007). World population prospects: The 2006 revision. Highlights. Working Paper No. ESA/P/WP.202. United Nations Department of Economic and Social Affairs, Population Division, www.un.org/esa/population/publications/wpp2006/WPP2006_Highlights_rev.pdf.

———. (2010). World urbanization prospects: The 2009 revision: Highlights. Department of Economic and Social Affairs, Population Division, http://esa.un.org/unpd/wup/Documents/WUP2009_Highlights_Final.pdf.

United Nations Development Programme (UNDP). (2013). Gender inequality index. Human Development Report Office, Nov. 15, 2013, http://hdr.undp.org/en/content/gender-inequality-index.

United Nations Environment Programme (UNEP). (2011). Decoupling natural resource use and environmental impacts from economic growth: A Report of the Working Group on Decoupling to the International Resource Panel: M. Fischer-Kowalski, M. Swilling, E. U. von Weizsäcker, Y. Ren, Y. Moriguchi, W. Crane, F. Krausmann, N. Eisenmenger, S. Giljum, P. Hennicke, P. Romero Lankao, A. Siriban Manalang, and S. Sewerin, www.unep.org/resourcepanel/decoupling/files/pdf/decoupling_report_english.pdf.

US Department of Commerce, US Census Bureau. (2011). Selected measures of household income dispersion: 1967 to 2010, https://www.census.gov/hhes/www/income/data/historical/inequality/IE-1.pdf.

———. (1949). Historical statistics of the United States, 1789–1945, www2.census.gov/prod2/statcomp/documents/HistoricalStatisticsoftheUnitedStates1789-1945.pdf.

US Department of Health and Human Services. (2014). Annual update of the HHS Poverty Guidelines. *Federal Register*, Jan. 22, 2014, https://www.federalregister.gov/articles/2014/01/22/2014-01303/annual-update-of-the-hhs-poverty-guidelines.

US Energy Information Administration. (2010). International energy statistics, www.eia.gov/cfapps/ipdbproject/iedindex3.fm?tid=90&pid=45&aid=8&cid=regions&syid=2006&eyid=2010&unit=MMTCD.

———. (2014a). International energy statistics, www.eia.gov/cfapps/ipdbproject/iedindex3.cfm?tid=5&pid=5&aid=2&cid=CG5,&syid=2009&eyid=2013&unit=TBPD.

———. (2014b). Electricity monthly update with data for September 2014, Nov. 25, 2014, www.eia.gov/electricity/monthly/update/.

Van Alstyne, Marshall, and Erik Brynjolfsson. (2005). Global village or cyber-Balkans? Modeling and measuring the integration of electronic communities. *Management Science* 51:(6):851–868, http://pubsonline.informs.org/doi/abs/10.1287/mnsc.1050.0363.

Veeraraghavan, Rajesh. (2013). Dealing with the digital panopticon: The use and subversion of ICT in an Indian Bureaucracy. Pp. 248–255 in *International Conference on Information and Communication Technologies and Development (ICTD2013)*, http://dx.doi.org/10.1145/2516604.2516631.

Veeraraghavan, Rajesh, Bharathi Pitti, Gauravdeep Singh, Greg Smith, Brian Meyers, and Kentaro Toyama. (2005). Towards accurate measurement of computer usage in a rural kiosk. In *Third International Conference on Innovative Applications of Information Technology for Developing World – Asian Applied Computing Conference*, Nepal, www.msr-waypoint.net/pubs/80531/KioskLogging.pdf.

Veeraraghavan, Rajesh, Gauravdeep Singh, Kentaro Toyama, and Deepak Menon. (2006). Kiosk usage measurement using a software logging tool. Pp. 317–324 in *International Conference on Information and Communication Technologies and Development (ICTD2006)*, http://research.microsoft.com/en-us/um/india/projects/kiosktool/rajesh_vibelog_berkeley.pdf.

Veeraraghavan, Rajesh, Naga Yasodhar, and Kentaro Toyama. (2009). Warana unwired: Mobile phones replacing PCs in a rural sugarcane cooperative. *Information Technologies and International Development* 5(1):81–95, http://itidjournal.org/itid/article/view/327/150.

Venkatesh, Sudhir. (2008). *Gang Leader for a Day: A Rogue Sociologist Crosses the Line*. Allen Lane.

Viola, Paul, and Michael Jones. (2001). Rapid object detection using a boosted cascade of simple features. In *Proceedings of the Conference on Computer Vision and Pattern Recognition*, http://dx.doi.org/10.1109/CVPR.2001.990517.

Vornovytskyy, Marina, Alfred Gottschalck, and Adam Smith. (2011). Household debt in the U.S.: 2000 to 2011. US Census Bureau, www.census.gov/people/wealth/files/Debt%20Highlights%202011.pdf.

Wahba, Mahmoud A., and Lawrence G. Bridwell. (1976). Maslow reconsidered: A review of research on the need hierarchy theory. *Organizational Behavior and Human Performance* 15:212–240, www.sciencedirect.com/science/article/pii/0030507376900386.

Wai, J., D. Lubinski, C. P. Benbow, and J. H. Steiger. (2010). Accomplishment in science, technology, engineering, and mathematics (STEM) and its relation to STEM educational dose: A 25-year longitudinal study. *Journal of*

Educational Psychology, Advance online publication, http://psycnet.apa.org/psycinfo/2010-19348-001.

Walensky, Rochelle P., and Daniel R. Kuritzkes. (2010). The impact of The President's Emergency Plan for AIDS Relief (PEPfAR) beyond HIV and why it remains essential. *Clinical Infectious Diseases* 50(2):272–275, http://cid.oxfordjournals.org/content/50/2/272.full.

Wallis, John Joseph. (2006). The concept of systematic corruption in American history. In Edward L. Glaeser and Claudia Goldin, eds., *Corruption and Reform: Lessons from America's Economic History*. University of Chicago Press.

Wall Street Journal. (2009). Sarkozy adds to calls for GDP alternative. *Wall Street Journal* blog, Sept. 14, 2009, http://blogs.wsj.com/economics/2009/09/14/sarkozy-adds-to-calls-for-gdp-alternative/.

Warschauer, Mark. (2003). Demystifying the digital divide. *Scientific American* 289(2):42–47, www.scientificamerican.com/article/demystifying-the-digital/.

———. (2006). *Laptops and Literacy: Learning in the Wireless Classroom*. Teachers College Press.

Warschauer, Mark, Michele Knobel, and LeeAnn Stone. (2004). Technology and equity in schooling: Deconstructing the digital divide. *Educational Policy* 18(4):562–588, http://epx.sagepub.com/content/18/4/562.short.

Watson, Tony. (2008). *Sociology, Work and Industry*, 5th ed. Routledge.

Weber, Max. (1904 [1976]). *The Protestant Ethic and the Spirit of Capitalism*. Talcott Parsons, trans. Charles Scribner's Sons.

———. (1915 [1951]). *The Religion of China*. Hans H. Gerth, trans. Free Press.

———. (1916 [1958]). *The Religion of India*. Hans H. Gerth and Don Martindale, trans. Anima Publications.

Weiner, Myron, ed. (1966). *Modernization: The Dynamics of Growth*. Basic Books.

Weir, Hugh. (1922). The story of the motion picture. *McClure's* 54(9):85–89.

White, Chapin. (2007). TRENDS: Health care spending growth: How different is the United States from the Rest of the OECD? *Health Affairs* 26:1154–1161, http://content.healthaffairs.org/content/26/1/154.full.

Wikipedia. (n.d.). Http://www.wikipedia.org.

Wills, Garry. (2002). *James Madison: The American Presidents Series: The 4th President, 1809–1817*. Henry Holt and Company.

Wilson, Edward O. (2012). *The Social Conquest of Earth*. Liveright.

Wood, Daniel B. (2013). An iPad for every student? What Los Angeles School District is thinking. *Christian Science Monitor*, Aug. 28, 2013, www.csmonitor.com/USA/Education/2013/0828/An-iPad-for-every-student-What-Los-Angeles-school-district-is-thinking.

World Bank. (2012a). GDP per capita (current US$), http://data.worldbank.org/indicator/NY.GDP.PCAP.CD/countries.
———. (2012b). GDP per capita (current US$), http://databank.worldbank.org/data/views/reports/tableview.aspx (tab set to India).
———. (2012c). World development indicators and global finance development database, http://databank.worldbank.org/data/home.aspx.
World Economic Forum. (2013). The global gender gap report 2013, www3.weforum.org/docs/WEF_GenderGap_Report_2013.pdf.
World Health Organization. (2011). Annual report 2011. Global Polio Eradication Initiative, www.polioeradication.org/Portals/0/Document/AnnualReport/AR2011/GPEI_AR2011_A4_EN.pdf.
———. (2014). Ambient (outdoor) air pollution in cities database 2014, www.who.int/phe/health_topics/outdoorair/databases/cities/en/.
World Values Survey. (2005). WVS 2005–2006 wave, OECD-split version (Ballot A), www.worldvaluessurvey.org/wvs/articles/folder_published/survey_2005/files/WVSQuest_SplitVers_OECD_Aballot.pdf.
———. (n.d.). Findings and insights, www.worldvaluessurvey.org/WVSContents.jsp.
World Wide Web Foundation. (n.d.). Connecting people. Empowering humanity, www.webfoundation.org/about/.
Wortham, Jenna. (2011). Feel like a wallflower? Maybe it's your Facebook wall. *New York Times*, April 10, 2011, www.nytimes.com/2011/04/10/business/10ping.html.
Wright, Robert. (2000). Nonzero: The logic of human destiny. Vintage.
Wydick, Bruce, Elizabeth Katz, and Brendan Janet. (2014). Do in-kind transfers damage local markets? The case of TOMS shoe donations in El Salvador. *Journal of Development Effectiveness* 6(3):249–267, www.tandfonline.com/doi/abs/10.1080/19439342.2014.919012.
Yang, Shih-Ying. (2001). Conceptions of wisdom among Taiwanese Chinese. *Journal of Cross-Cultural Psychology* 32:662–680.
Yaqoob, Tahira, and Laura Collins. (2011). Wael Ghonim: The voice of a generation. *The National*, Feb. 12, 2011, www.thenational.ae/news/world/africa/wael-ghonim-the-voice-of-a-generation.
Yunus, Muhammad. (1999). *Banker to the Poor: Micro-Lending and the Battle Against World Poverty*. PublicAffairs.
———. (2007). *Creating a World Without Poverty: Social Business and the Future of Capitalism*. PublicAffairs.
———. (2011). Sacrificing microcredit for megaprofits. *New York Times*, Jan. 14, 2011, www.nytimes.com/2011/01/15/opinion/15yunus.html.

Zachary, Lois J. (2012). *The Mentor's Guide: Facilitating Effective Learning Relationships*. Jossey-Bass.

Zuckerberg, Mark. (2014). Facebook post, March 27, 2014, https://www.facebook.com/zuck/posts/10101322049893211.

Kentaro Toyama is W. K. Kellogg Chair Associate Professor at the University of Michigan's School of Information and a Fellow of the Dalai Lama Center for Ethics and Transformative Values at the Massachusetts Institute of Technology. Until 2009, he was assistant managing director of Microsoft Research India, which he cofounded in 2005, and where he researched how the world's poorer communities interact with electronic technology and invented new ways for technology to support their socioeconomic development. He lives in Ann Arbor.

PublicAffairs is a publishing house founded in 1997. It is a tribute to the standards, values, and flair of three persons who have served as mentors to countless reporters, writers, editors, and book people of all kinds, including me.

I. F. STONE, proprietor of *I. F. Stone's Weekly*, combined a commitment to the First Amendment with entrepreneurial zeal and reporting skill and became one of the great independent journalists in American history. At the age of eighty, Izzy published *The Trial of Socrates*, which was a national bestseller. He wrote the book after he taught himself ancient Greek.

BENJAMIN C. BRADLEE was for nearly thirty years the charismatic editorial leader of *The Washington Post*. It was Ben who gave the *Post* the range and courage to pursue such historic issues as Watergate. He supported his reporters with a tenacity that made them fearless and it is no accident that so many became authors of influential, best-selling books.

ROBERT L. BERNSTEIN, the chief executive of Random House for more than a quarter century, guided one of the nation's premier publishing houses. Bob was personally responsible for many books of political dissent and argument that challenged tyranny around the globe. He is also the founder and longtime chair of Human Rights Watch, one of the most respected human rights organizations in the world.

• • •

For fifty years, the banner of Public Affairs Press was carried by its owner Morris B. Schnapper, who published Gandhi, Nasser, Toynbee, Truman, and about 1,500 other authors. In 1983, Schnapper was described by *The Washington Post* as "a redoubtable gadfly." His legacy will endure in the books to come.

Peter Osnos, *Founder and Editor-at-Large*